Coral Reef Geomorphology

Coastal Morphology and Research

Series Editor: Eric C. F. Bird

CORAL REEF GEOMORPHOLOGY

André Guilcher

Coral Reef Geomorphology

André Guilcher
Professor Emeritus
University of Brest

JOHN WILEY & SONS
Chichester · New York · Brisbane · Toronto · Singapore

Library of Congress Cataloging-in-Publication Data

Guilcher, André.
Coral reef geomorphology.

(Coastal morphology and research)
1. Coral reefs and islands. I. Title. II. Series.
GB461.G85 1988 551.4′24 87-25422
ISBN 0 471 91755 9

British Library Cataloguing in Publication Data

Guilcher, André
Coral reef geomorphology.—(Coastal Morphology and research).
1. Coral reefs and islands
I. Title II. Series
551.4′24 GB461

ISBN 0 471 91755 9

Typeset by Acorn Bookwork, Salisbury, Wiltshire.
Printed and bound by The Bath Press, Bath, Avon.

Contents

Foreword

One of the aims of the Coastal Morphology and Research series is to present up-to-date accounts of coastal topics, especially those on which there is a substantial literature published in languages other than English. Coral reef geomorphology is one topic which has a literature in various languages, and over the past 35 years Professor André Guilcher, of the University of Western Brittany, has been prominent among those who have written in French on coral reefs.

I knew of Professor Guilcher in my student days, when I used his *Morphologie Littorale et Sous-marine* (1954), which became well known as a textbook in its English version, *Coastal and Submarine Morphology* (1958), and we had corresponded for several years before I met him, in Canberra, just before his first field session in the Society Islands in 1963. We have maintained a dialogue over exchanged reprints in the succeeding quarter century, and have worked together in the field in various parts of the world, notably Kenya and Saudi Arabia, as well as in his native Brittany.

The *doyen* of French coastal geomorphologists and oceanographers, Professor Guilcher's first discussion of the forms and origins of coral reefs was published in *L'Information Géographique* in 1950. He was a member of the French oceanographic expedition to the Red Sea, directed by Commander Jacques Cousteau on the *Calypso* in December 1951–January 1952, which included work on the coral reefs between Jeddah and the Farsan Bank. His subsequent research on coral reefs has taken him from 1954 to 1985 to Madagascar, Mayotte Island, New Caledonia, the Society and Tuamotu Islands, the Solomon Islands, Kiribati, Vanuatu, Florida, the Sinai Peninsula, Lord Howe Island in the Tasman Sea, Guadeloupe Island, Haiti, Paraiba in Brazil, and Kenya, and he has published fifty-three papers on coral reef geomorphology. Many of these dealt with reefs in areas that have received little attention from English-speaking geomorphologists, who have worked primarily in the Caribbean, the central and eastern Indian Ocean, Micronesia, and the Great Barrier Reef in Australia. In recent years successive international coral reef conferences have widened the scope of these studies, and there is now a need for a comprehensive account of coral reef geomorphology, pinpointing the various unsolved problems. I believe that the present volume will achieve this, and I hope that it will stimulate interest in coral reefs among students and young researchers, thereby generating further activity in a field of research that was initiated by the observations of Charles Darwin a century and a half ago.

Eric Bird

Introduction

1. SCOPE OF THE BOOK

The scope of the book is geomorphological rather than biological. It does not list or describe in detail organisms found at present and in the past on and in coral reefs; it will give only basic data in this respect. It considers primarily the shape and types of reefs, the way in which they are now being built, how they have grown in the past to attain their present form, the contrasts observed in reef structure and development in different parts of the oceans, the damage caused in them by human activities and the requirements for reef protection and recovery.

Background information on the geology and ecology of coral reefs has been summarized in a review by Stoddart (1969a), with an imposing bibliography, and in more recent reviews of the problems of metabolism, calcification and carbon production in coral reefs (Kinsey, 1985; Gladfelter, 1985). The reader wishing to learn about the biological aspects of reef growth must refer to these reviews. The literature on coral reef geomorphology has expanded rapidly by way of papers published in *Atoll Research Bulletin*; the journal *Coral Reefs* issued from Heidelberg since 1982; the Proceedings of five international symposia or congresses on Coral Reefs (Mandapam Camp, India, 1969, published 1972; Great Barrier Reef of Queensland, 1973, published in Brisbane, 1974; Miami, USA, 1977; Manila, Philippines, 1981; Tahiti, French Polynesia, 1985); and Hopley's fine study *Geomorphology of the Great Barrier Reef* (1982), which includes discussions of general problems and areas lying outside the Great Barrier Reef. The Japanese book by Tayama, *Coral Reefs in the South Seas*, 1952, will be extremely valuable for those who can find it. Extensive bibliographies may be found in Davis, 1928, in Guilcher, 1950, in Selected Bibliography . . . 1958, in Milliman, 1965, and in *Coral Reef Newsletter*. This book has been written in 1986 and 1987.

2. DEFINITION OF CORAL REEFS

The indispensable element in structures deserving the appellation of coral reefs is the presence of colonies of scleractinian hermatypic corals or madrepores (i.e. stony corals, from the Greek σκληρός, hard, and typically built-up reefs, from ερμα, rocky basement; see however the discussion in Chapter one, 7). Organisms used as raw products for jewellery, and known as pink coral (*Corallium rubrum*), black coral (*Antipathes grandis*, *A. dichotoma*), gold coral (*Parazoanthus*) and bamboo coral (*Keratoisis nuda*), live for a large part in places outside the tropical belt, as the Mediterranean Sea and the southern fjords of New Zealand, and at rather great depths in tropical areas (Hawaii, Taiwan, etc) (Grigg, 1974; Grange, 1985), but do not thrive currently in true coral reefs. The latter are communities of various species among which the madrepores, although always present, are generally not the most numerous. It is rather exceptional to find reefs which, outside their outer slopes, include more than fifty per cent of living madrepores; and the presence of many dead colonies beside a few living ones does not mean necessarily that the reef as a whole is dying, although such a decay can readily happen.

More comments and details on the definition of coral reefs will be found in Chapter 1, after the reef ecology has been considered.

The two most important genera among the scleractinian hermatypic corals are *Acropora*, or branching corals, and *Porites*, or massive corals; both of which include a large number of species. Other important genera are *Fungia*, *Montipora*, *Pocillopora*, *Leptoseris*, and *Millepora*. A great diversity of species over small areas characterizes the most thriving coral reefs and especially their slopes, and makes them resemble gardens with assemblages of variegated flowers (Fig. 1).

Fig. 1 Living corals exposed at a low spring tide in Ramanetaka Bay, west coast of Madagascar. Tidal range: 4 metres at spring tides (Photo A. Guilcher)

But coral reefs form symbiotic associations in which many other living organisms contribute and shape a common structure. These organisms include algae, foraminifera, crustaceans, gastropods, lamellibranchs, echinoderms, holothurians, and fishes. Alcyonariae, or soft corals, are frequent. Algae, which are almost as important as corals in reef building, sometimes form a hard pavement, as with *Porolithon*; others, such as the various Halimedae, are soft, and contribute by their platelets, when they die, to the loose sediments scattered over the reef. These sediments are of various initial sizes, consisting also of tests of foraminifera, faeces of invertebrates, products of the grazing of corals by fishes, and are modified by wave action. They partly fill cavities in the reef structure, and are often cemented by algal coatings. Another type of cementation results in the formation of beach rock within beaches built up behind coral reefs. All the organisms, including the fishes, must be considered components of hermatypic reefs. The upper limit of reef growth is the level of ordinary low tides, a very precise level which is a datum for considering levels of fossil, emerged reefs. The corals cannot survive emergence for more than two hours during the lowest spring tides (Fig. 1). On reef surfaces, mechanical abrasion and bioerosion occur at the same time as construction. More or less similar combinations of physical processes and biological or biochemical associations are found in coastal geomorphogenesis outside of coral reefs; but the existence of growing madrepores in this environment makes it distinctive.

When lying far out in the oceans, the reefs are exclusively made of calcium carbonate, apart from a few siliceous spicules of sponges, diatoms, or very fine particles of dust carried by winds; but many reefs lie close to continents, subcontinents such as New Guinea and Madagascar, or oceanic volcanoes, and their sedimentological framework is then strongly influenced by terrigenous inputs.

The reefs never form a completely compact and massive structure. Numerous cavities remain in the framework, both at and near the surface and at depth. If the reef emerges above sea-level for some reason, either tectonic uplift or lowering of the sea, its geomorphological evolution continues in spite of the death of its marine components. Some of the holes can be filled by continental sedimentation, while others can be created or enlarged by karstic processes. Except where a reef has been strongly folded and completely deformed, it is still a coral reef landform, and the study of these raised coral features remains within the discipline of geomorphology.

3. DISCOVERY AND EXPLORATION

The Portuguese, in their travels to India by way of South Africa, were around 1500–1510 the European discoverers of coral reefs located far from the coasts in the Indian Ocean. The Arabs, during their previous voyages between Arabia, Africa and India, had certainly seen many of them, but these structures remained generally unnamed, so far as we know, and uninhabited, the main exceptions being the reefs in the vicinity of India in the east and those close to Africa and Madagascar in the west. The names given by the Portuguese sailors for a number of these reefs, or mountainous islands surrounded by reefs, have persisted, as in Bassas da India (Baixo de Judia), Saya de Malha, Rodriguez, Peros Banhos, and Diego Garcia which is perhaps the finest atoll in that ocean. Later on, the French, the Dutch and the British added their contributions to the toponymy—Tromelin, Coetivy, Mauritius, Farquhar, Blenheim—and to the settlements.

Man has lived much earlier on the Pacific coral islands, in spite of their remote situation, since the Polynesians are thought to have arrived in the middle of that ocean about the beginning of the Christian era, and to have reached Hawaii seven or eight centuries later. The Spaniards (see Atlas Okeanov) were the first Europeans to discover parts of this coral realm, with Mendana who landed in the Tuvalu, Santa Cruz and Solomon Islands in 1568 and 1595; and Abel Tasman from the Netherlands saw the Fiji and the Tongan Islands in 1643; but it was not before the second half of the eighteenth century that scientific expeditions led by the British (Wallis and Cook) and the French (Bougainville and Lapérouse) began to give a fairly good knowledge of a large number of Pacific reefs. The Russian contributions (Kotzebue, Bellingshausen of German Baltic origin, 1816–1820) in the Marshalls and the Tuamotus, were also very valuable. Since that time this ocean has provided the best information on coral reefs, particularly the Tuamotu and Society Islands, which gave Charles Darwin in 1835 the first and decisive opportunity to check his famous hypothesis on the genesis of their different frameworks. In his book (1842, p. 99), he chose the 'encircled island of Bolabola' as the best example of a barrier reef.

In the Atlantic, chance took Columbus to land, on his first voyage in 1492, in the Bahamas, which can be more or less grouped with coral reefs, although they are of a quite special nature; he also discovered in the same year Cuba and Hispaniola, which have typical coral reefs. The same explorer found, in the following years, the Leeward Islands from Anegada to Dominica, Puerto Rico and Jamaica, all of them bearing hermatypic corals, so that the reefs of the Caribbean area as a whole began to be known by Europeans before the end of the fifteenth century. By 1532, the reefs bordering the continental coast of the Caribbean Sea, which are very numerous in Belize, had also been explored by the Spaniards. These findings, however, were not so important from a scientific point of view as those made in the two other oceans, since the Caribbean coral reefs, although including very interesting structures and even examples of the barrier and atoll types, do not occupy a place comparable to their Pacific and Indian counterparts. On the other hand, the Portuguese made a major contribution to the discovery of the Atlantic reefs, since they were in 1500, with Cabral, the first probable visitors to the east coast of Brazil, where the name Abrolhos, applied to the most typical reefs in that area, bear testimony to these Lusitanian findings.

In spite of their location closer to Europe, the coral reefs growing along and off the sides of the Red Sea are still among the least known in the world, especially in their central and southern parts, because of their intricate pattern, their situation in a sea that is difficult for navigation, and the political situation in that area. Investigations so far carried out by German, French, British, Israeli, Arabian, and American teams seem to indicate that the Red Sea is perhaps the region where the most important discoveries in reef geomorphology are still to be made.

CHAPTER ONE

Distribution and ecology

1. DISTRIBUTION

The area in which contemporaneous hermatypic coral reefs thrive is mostly in the intertropical Indo-Pacific region. They are represented (Fig. 2) by more than 30 genera from 15°S (Kenya, North Madagascar, Northern Territory and Queensland in Australia, New Caledonia, Society Islands) to 10–20°N (Northern Red Sea, the Maldives, the Andamans, the Philippines, the Marshalls), as far as, approximately, 150–160°W. A second area, not so prolific but still important, with more than 20 genera, is found around islands and along the continental coast bordering the Caribbean Sea. In the Pacific as in the Atlantic, the number of genera decreases from west to east. In the Pacific, 35 hermatypic genera are recorded in the Society Islands, 30 in the Tuamutus, and 17 in the Marquesas, where they have not built true reefs (Chevalier, 1979). The Pacific coast of the Americas is poor in reefs, having only small, though sometimes fine, coral colonies in Baja California and off Central America (Durham, 1947), e.g. on Cocos Island off Costa Rica (Bakus, 1975), and on the northeastern side of the Pearl Islands, Gulf of Panama (Glynn and Stewart, 1973). In the Atlantic, outside the Caribbean, they are comparatively poor on the east coast of Brazil, the best ones being on the Abrolhos (Laborel, 1969). On the west coast of Africa (Laborel, 1974, and *in litt.*), true scleractinian hermatypic corals are practically non-existent: small coral colonies no more than a few metres wide and high are scattered around the isles of inner Gulf of Guinea, on Annobon for example. Others, generally smaller, exist in sheltered 'coral oases' off Equatorial Guinea, Ghana, and around the Cape Verde Islands. The only eastern coast of an ocean where true scler-

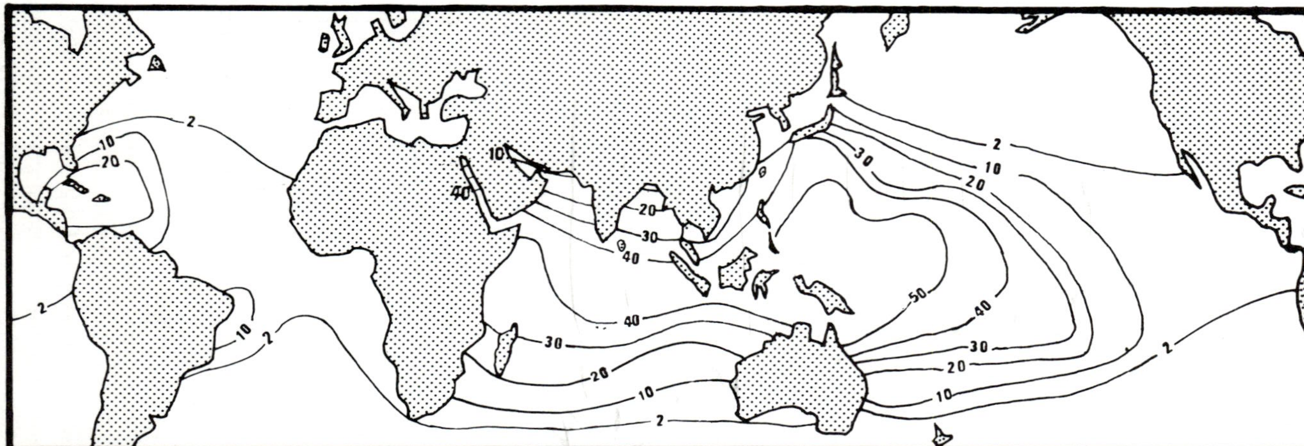

Fig. 2 Distribution of hermatypic coral genera. From Stoddart (1969), with modifications after Laborel (1969) (Brazil) and Chevalier (1979) (South Pacific)

actinian, reef-building corals are extensive is the coast of Western Australia, where a vigorous growth of corals occurs at 28–29°S in Houtman's Abrolhos (not to be confused with the Brazilian Abrolhos).

The northern limit of hermatypic reefs in the Atlantic lies in Florida and in the Bermudas. Local reefs are found in the southwest and south of the Persian or Arabian Gulf in Bahrain, Qatar and the bay east of Qatar, and even in Kuwait at a latitude of 29°N, where they are smaller and poorer than those living in the Gulf of Elat or Aqaba, in the northernmost part of the Red Sea (Purser and Seibold, 1973; Downing, 1985). Coral distributions on the Iranian coast are not precisely known. Reefs are absent, or do not thrive, in the northern part of the Bay of Bengal and in the inner Gulf of Siam, or along the inner reaches of the coast of China; but reef-building corals are well represented in the Ryukyu Islands, and exist to some extent in the southern part of the main Japanese Islands. Reefs live in Hawaii as far north as Kure Island (29°N). In the southwestern Indian Ocean, reefs extend with a moderate growth to South Moçambique (26°S) and to northernmost Natal, but they are not as well developed as in Western Australia. The southern boundary in the Pacific (and in the world) lies at Lord Howe Island, Tasman Sea, 31°30′S, that is at approximately the same latitude as the Bermudas in the northern hemisphere.

Considering the areas covered by coral reefs and associated communities living at depths of less than 30 metres, Smith (1978) has calculated that 30 per cent are located in the adjacent seas between Indonesia, Northern Australia, the Philippines and continental Asia; 25 per cent in the Pacific Ocean; 30 per cent in the Indian Ocean, Red Sea and Arabian or Persian Gulf; 14 per cent in the Caribbean Sea and North Atlantic; and 1 per cent in the South Atlantic.

2. EXPLANATIONS: SEA TEMPERATURE

The basic explanation of the reef distribution is sea temperature (Atlas Okeanov, 1974–1977). Although some species can tolerate temperatures down to 15 °C, and even less, and up to 36 °C, both for brief lapses of time, the limits for normal growth are 17–18 °C and 33–34 °C. The main obstacle to coral growth in low latitudes along the east coasts of the oceans (i.e. west coasts of continents) is the presence of cool water, with a temperature lower than 18 °C for at least part of the year, and sometimes much lower (Fig. 3). This is caused by trade winds, blowing parallel to the coast in both hemispheres, and as a result of the Coriolis force producing a current to the northwest in the southern hemisphere and the southwest in the northern hemisphere, away from the coasts, resulting in an upwelling of cool water (13–14 °C on average) from 100–150 metres depth. These cool currents are referred to as the California, Humboldt (or Peru), Canary, and Benguela currents in the Pacific and the Atlantic. They are subject to fluctuations, and are sometimes displaced by warmer currents, such as the famous El Niño off Peru, and probably a more or less equivalent current in South Angola

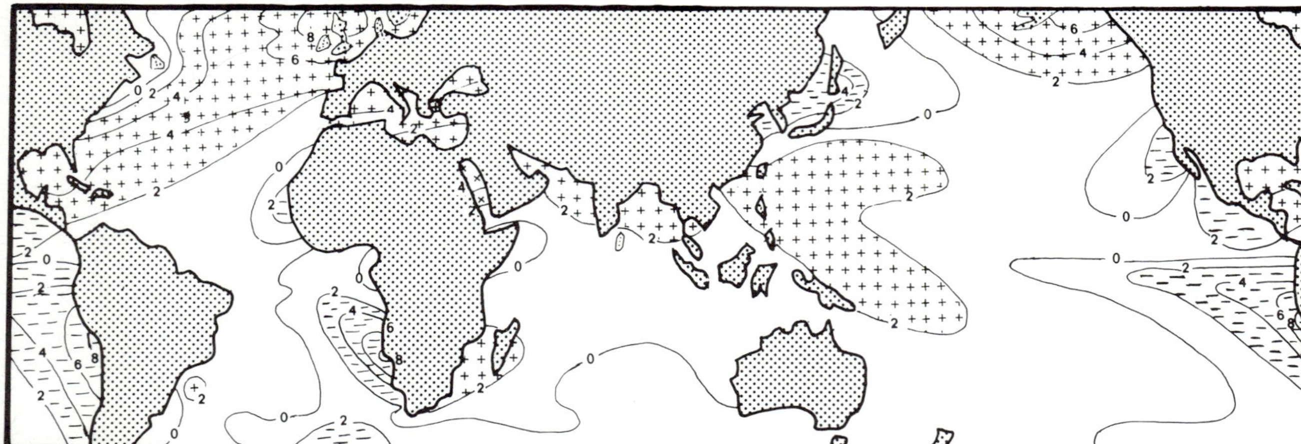

Fig. 3 Isanomals of surface temperature referred to the values which would exist on a globe entirely covered by the ocean. After Dietrich (1950). Values in C°. Explains contrasts in distribution of corals on eastern and western coasts under low latitudes (cf. Fig. 2)

(Guilcher, 1982b; confirmed by Hirst and Hastenrath, 1983); but the presence of cool water during the larger part of the year is sufficient to prevent the formation of coral reefs. The exceptional occurrence of small reefs on the northeastern side of the Pearl Islands, Gulf of Panama, results from a local absence of upwelling, whereas on the southwestern side coral growth is inhibited by cool upwelling caused by the northeasterly trade winds (Glynn and Stewart, 1973). Coral growth is locally hindered by summer upwelling on part of the Somalian coast.

The difference in coral growth between the Persian or Arabian Gulf and the Gulf of Elat or Aqaba is partly explained by the shamal, a cold northwesterly wind blowing in winter over the northern part of the Persian Gulf, so that the temperature in that area is lowered to 15 °C (Emery, 1956a), and occasionally to 13.2 °C on the Kuwait reefs (Downing, 1985). For another explanation related to salinity in the Persian Gulf, see below.

Conversely, the high latitude of the Lord Howe Island reefs, and the adjacent Middleton and Elizabeth atolls (29 to 31°S) results from the east Australian warm current, which flows from the Coral Sea to the Southern Ocean (sea temperature: 18 to 24 °C according to month). The Bermuda reefs receive warm water flowing along the right-hand side of the Gulf Stream (19 to 27 °C), and the Ryukyu and other southern Japanese islands are favoured by the warm Kuro Shio current (20 to 28 °C). At Kure in the Hawaiian Islands, the thermal range is 18 to 27 °C. The exceptional occurrence of coral reefs at the Houtman's Abrolhos, Western Australia, results from the equally anomalous Leeuwin current, flowing polewards along this eastern oceanic coast, and carrying warm equatorial Pacific water (never less than 20 °C, a figure which occurs in September) around northwestern Australia.

It is, however, difficult to explain in terms of sea temperature the absence of well-developed reefs in the Gulf of Guinea. It is true that at Pointe Noire, Republic of Congo, there is the anomaly of an annual thermal range of 7 °C at latitude 5°S, but the temperature does not fall below 17–18 °C in July–August, and attains this figure only for one month. This does not seem sufficient to prevent coral growth, and farther north, in the Gulf of Guinea proper, the sea temperature remains higher throughout the year.

3. SALINITY

Another factor in reef growth is salinity. Corals are rather tolerant in this respect, accepting a range from 30 to 38 parts per thousand, and even 40–41 parts per thousand in the Red Sea. An even higher salinity can, however, limit coral growth, as in the inner, shallow reaches of the bays on both sides of Qatar Peninsula, Arabian Gulf: in these overheated areas, salinities rise to at least 50–60 parts per thousand in summer, and corals are restricted to the more open parts of bays where salinity is diminished by circulation of sea water (Purser and Seibold, 1973). Similar conditions are found in Shark Bay, Western Australia. Conversely, it might be that the large quantities of freshwater delivered by the Congo, Ogooué and Niger rivers play a part in the paucity of reefs on neighbouring coasts, and the same may be true along the coast of Liberia, where a permanent deficit of salt occurs (Guilcher, 1968).

4. TURBIDITY

Water turbidity is even more important in preventing reef growth, especially in the Indian Ocean and the South China Sea, where coral reefs are scarce and scattered, in spite of a suitable temperature. The same is true on the coast of Brazil (see Chapter 4, section 7, for more detail). Some corals, like *Fungia*, are able to prevent accretion of silt particles by movement of their cilia, but others cannot do this. The result is thus complex. Corals exist on many parts of the coast of Indonesia on muddy substrates, in the vicinity of river mouths which supply large quantities of suspended mud, as in the Bay of Jakarta, around Celebes and Sumatra, on the coast of Ceram, and in the Aroe Islands (Kuenen, 1933; Umbgrove, 1947; Verstappen, 1953). The same is observed at Madagascar, off the mouth of the River Loza (Guilcher, 1956). On the eastern shelf of Nicaragua (Roberts and Murray, 1983), extremely high rainfall leads to rapid erosion of volcanic rocks, and results in the supply of 15 times more sediment per unit length of coast than along the eastern shore of the United States: nevertheless, the turbidity is restricted to a narrow band of southward-moving coastal water, and the remainder of the shelf is a carbonate province, with active growth of coral.

In general, however, turbidity is an explanation of the absence or paucity of coral reefs, especially in South and East Asia, Indonesia and Melanesia, where rivers deliver 70 per cent of the world supply of sediments to the oceans (e.g. Ganges–Brahmaputra: 1670 million tonnes/year; Yangtse/Chan Jiang: 480; Irrawaddy: 265; Mekong: 160: Taiwan rivers 185; Purari and Fly in Papua New Guinea: 110), while none of the African rivers, even the Congo (Zaire), the Niger, the Limpopo and the Zambezi contribute as much as 45 million tons/year (Milliman and Meade, 1983). Although the effects are complex, water turbidity is probably as important as sea temperature in the distribution of reefs.

The poor development of coral reefs in the Marquesas Islands in the Southeast Pacific is a difficult problem, often discussed without any definite conclusions. There are only very scattered and embryonic fringing reefs around these high volcanic islands. Recent research (Brousse *et al.*, 1978) concluded that this absence could be due to a combination of several factors: a paucity of available hermatypic species, perhaps local upwelling of cool waters, locally turbid waters, and the presence of competing organisms. Still more recently, Rougerie and Wauthy (in press) mention platforms drowned at 90 metres and thought to be former barrier reefs.

5. ECOLOGY

Hermatypic corals owe their growth to a symbiotic association with unicellular algae, the Zooxanthellae. The classic discoveries in this respect are principally due to Yonge (1930–31, 1972), who observed, during his studies deriving from the first great expedition to the Great Barrier Reef of Australia, that corals feed on zooplankton which is passed into their stomachs by the action of cilia, producing a constant flow of water; but that their respiration depends on the Zooxanthellae included in the cells of the inner covering of the animals. The corals receive their oxygen from the algae; the algae in turn receive nutrients and carbon dioxide from their hosts and, in addition, remove the coral waste products. The high production of calcium carbonate in coral reefs is thus associated with the presence of plant life, directly in the case of the algae living on the reefs (species of Rhodophyta, Chlorophyta and Phaeophyta), and indirectly by way of symbiotic Zooxanthellae. A consequence of this symbiosis is that such corals do not live in places where sunlight does not penetrate, because the Zooxanthellae, like other plants, require sunlight for their metabolism. Accordingly, the reef-building corals thrive mostly between ordinary low tide level and a depth of 25–30 metres; below this depth the number of species falls. Hermatypic corals reported at −115 metres in the Bahamas (Reed, 1985) are rare exceptions. Even at 30–40 metres, the massive colonies give place to thinly branching and leafy skeletal forms; corals with massive hemispherical heads in shallow water become flatter as depth increases, while branching species become thinner. The calcareous algae that contribute directly to reef growth (see above) also require sunlight, but some species are able to live at greater depths (see section 6 of this chapter).

Since the publication of Yonge's research, there have been controversies over the relationships between corals and Zooxanthellae. The discussion between 1955 and 1969 has been summarized, with references, by Stoddart (1969a, p. 446–449). In particular, Odum and Odum (1955) have discovered, in coral skeletons of the Marshall Islands, filamentous algae which led them to consider that the hermatypic corals were themselves primary producers, and have thus minimized the role of Zooxanthellae. Goreau (1963) contested these conclusions and showed clearly that Zooxanthellae play an important part in the calcification of corals. From a study of 7000 coral skeletons collected in many different regions of the tropical and subtropical seas, Weber (1974) concluded that 'the reef-building corals could not have evolved without first developing a symbiotic association with Zooxanthellae, and that large polyps and high polyp packing densities of the most modern hermatypic corals would not be possible without Zooxanthellae to remove waste metabolites that would otherwise be eliminated by slow diffusion through the coral tissue to seawater'. Again, Muscatine *et al.* (1985) found that in one of the major corals of the Red Sea more than 95 per cent of the net carbon fixed daily by the Zooxanthellae was translocated to the host animal tissue. On the other hand, adverse effects of endemic Zooxanthellae expulsions, causing coral bleaching and destruction of living reefs, have been described in the Florida Keys (Jaap, 1985), and in Queensland (Fisk and Done, 1985). Other recent studies (Chalker *et al*., 1987) have shown that algal photosynthesis enables the coral organism to deposit its limestone skeleton 2.5–5 times faster in the light than in the dark. Moonlight is also important: monthly periodicities have been observed and attributed to lunar influence. The role of the unicellular algae is thus fundamental in coral reef building.

The morphology of algal symbiosis with corals has been investigated in great detail (see Gladfelter's review, 1985). In some genera such as *Pocillopora* and *Montastrea*, the living tissue is quite superficial; in others such as *Acropora* and *Porites*, the tissues extends more deep, and ramifies through the skeleton. These differences have consequences in respiration of the corals and in the metabolism of the tissues.

Changes in temperature may influence the rate of both photosynthesis and respiration; but how these changes might affect the role of Zooxanthellae has not been determined. On the other hand, the metabolic adaptation of corals and their associated organisms to different hydrodynamic conditions is not well known.

Numerous studies of the rates of skeletal growth of corals in relation with various environmental factors have been reviewed in detail by Gladfelter (1985).

6. DEEP SEA CORAL PROBLEMS

So far only hermatypic corals, living near the surface in a manner determined by light requirements, have been considered. Other corals, described as ahermatypic, live in darkness at depths down to 6000 metres, at temperatures as low as −1.1 °C, and without Zooxanthellae (Teichert, 1958). In terms of the morphology and structure of the skeletal elements constituting those coral-like organisms, no distinction can be made between corals which do and do not possess Zooxanthellae (Weber, 1974). Hermatypes and ahermatypes may be grouped in the same families. But the deep-sea corals are smaller, and grow more slowly. Those growing off the Norwegian coast, and on the continental slope off France and the British Isles, are particularly well known (Fig. 4). Coral colonies have also been described in the Mediterranean Sea at several hundred metres depth: some are still living, some are not (thanatocoenoses) (Blanc *et al*., 1959). A recent survey by Cairns and Stanley (1981) enumerated fifteen major regions of ahermatypic coral development in the Atlantic and the Pacific, including newly-discovered areas in the sub-Antarctic South Pacific and off Chile, as well as the Gulf of Guinea, the sea-floor off Morocco, Florida, and New Zealand. In many places they grow as individual colonies. Off Norway *Lophelia prolifera* forms dendroid structures up to 0.6 metre in diameter, at depths of at least 600 to 900 metres; but

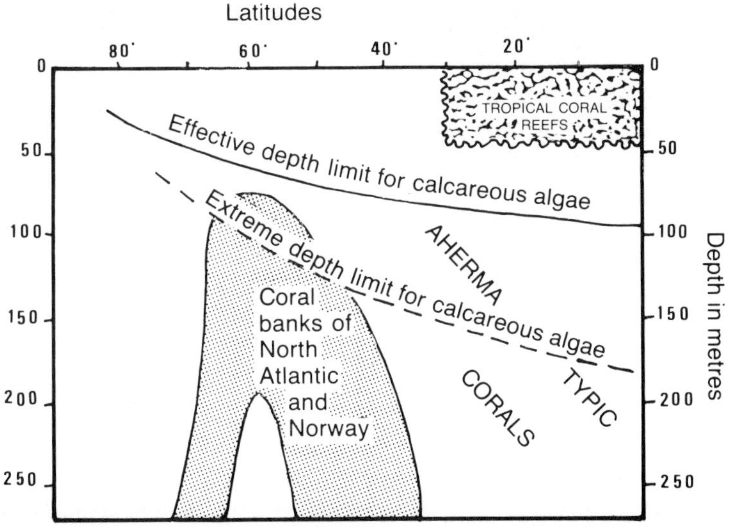

Fig. 4 Distribution of reef-building and hermatypic corals and of calcareous algae according to latitude. Deep coral banks in North Atlantic. Modified from Teichert (1958)

locally they form true reefs in the same area, so that the term 'ahermatypic' may not be etymologically suitable, although it seems generally accepted that their growth has been much slower. For example, Dons (1944) mentions a submarine bank off Norway, built by *Lophelia* and *Madrepora*, which covers an area of 1.5 × 2 kilometres and rises to 40 metres above the surrounding sea-floor. Another such bank reaches a height of 60 metres, and is thought to have formed within the last 4000 years. Some structures of this kind have been estimated to have grown between 0.75 and 1.5 centimetres annually (?). As with the common hermatypic corals living near the surface in tropical seas, these Norwegian corals are accompanied by biocoenoses that include sponges, coelenterates, worms, bryozoans, echinoderms, crustaceans, and molluscs, a total of 190 species. But there are no calcareous algae, which die out below 45–50 metres off Norway, whereas in tropical waters they can be found in some cases down to 150–180 metres according to Lemoine (1940), probably because of a better light penetration.

When these submarine reefs have built banks 40 to 60 metres high, such as those described by Dons (1944), a geomorphologist is tempted to consider them as true reefs: in fact, there are many fringing reefs or coral knolls in the tropical world which are not as big as this. If this is the case, the distinction between hermatypic and ahermatypic should perhaps be reviewed. If they were raised above the sea, they would be regarded as reefs; but Teichert (1958) reported that submarine reefs that have emerged near Oslo and Trondheim have been destroyed by subaerial denudation, so that this possibility seems theoretical.

7. A POSSIBLE REDEFINITION OF CORAL REEFS

Schuhmacher and Zibrowius (1985) have recently noted that the term hermatypic 'includes a confusing generalization of equating reef building' with the association of Zooxanthellae, while reef-building corals do not necessarily require such an association. Accordingly, they proposed the following definitions:

- I A. Zooxanthellate: living in symbiosis with dinoflagellate algae.
- I B. Azooxanthellate: without those symbionts. But some species include A and B individuals or colonies.
- II A. Constructional: forming a bioherm, i.e. an elevated durable carbonate structure in shallow or deep water; constructional forms are not necessarily hermatypic, but hermatypic forms are always constructional.
- II B. Non-constructional: not forming such a structure.
- III A. Hermatypic: significantly contributing to the framework of reefs, defined as persistent, positive topographic biogen structures, rising up to the surface of the sea, and characterized by their capability to resist hydrodynamic stress (however, such a definition by Schuhmacher *et al.* leaves out the possibility of deep-sea reefs, mentioned above). All hermatypic forms are constructional.
- III B. Ahermatypic: not significantly contributing to the framework of reefs. But the separation between III A and B is not, in Schuhmacher's opinion, clear cut.

8. CORALS IN THE GEOLOGICAL PAST

Many marine calcareous rocks outcropping on the continents are considered as 'reefs' by structural geologists and palaeobiologists, because they are coralliferous, i.e. partly or wholly made of reef-building corals. In terms of geomorphology two problems must be examined.

The first problem concerns their shape. Very often the original form of the reef has not been preserved, in which case they are not included in coral reef geomorphology. But in several instances the shape is still visible in the landscape, as in the Late Jurassic formation of the Caucasus North of Tbilisi (Bendukidze, 1977) (Fig. 5), where a complex described and illustrated consists of a wave-resistant barrier reef, with *in situ* hermatypic colonies and blocks eroded by the surf and then cemented in the reefs. The whole framework still stands in the topography above sediments deposited in a backing lagoonal environment. Such a fossil reef, although built during the Jurassic, is still a geomorphological feature. Another striking example is the palaeo-atoll of Jebel Bou Dahar, in the Liassic limestones of the Middle Atlas of Morocco (Du Dresnay, 1977), which has an elliptical rim of massive coral reef including *in situ* coral colonies, an outer slope with spurs and grooves (see Chapter two, 1.2, for this feature in modern reefs) and detrital blocks, and an inner area that was formerly a lagoon.

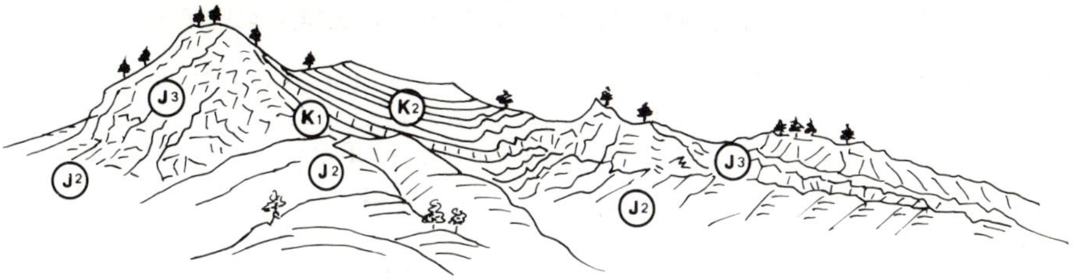

Fig. 5 Barrier reef in Jurassic limestone, as seen in the landscape of the Caucasus, north of Tbilisi, Georgian SSR. Redrawn after Bendukidze (1977). Trees give the scale. J2: Porphyritic Bajocian; J3: Oxfordian–Kimmeridgian hermatypic coral reef, built in surf environment as a barrier reef on left, passing into lagoonal facies on right; K1, K2: Lower Cretaceous marine complex

The second problem deals with the ecology of these old reefs, which is not simple, since corals have existed in the oceans since the Lower Cambrian, some 600 million years ago (e.g. Sedlak, 1977; Zamarreno and Debrenne, 1977). They progressively evolved during the Silurian and Devonian into organisms which built full-scale atolls and barrier reef complexes, but the data on these old reefs do not allow us to consider all of them *a priori* as sure indicators of either tropical environments or shallow water conditions. Indeed, several geologists have thought that some of them formed in deep water, such as the Cretaceous 'reefs' of Sweden and Denmark (Hadding, 1941). On the other hand, the Jurassic 'reefs' of England, which are generally thin and devoid of calcareous algae, contain oolites which could be an indication of a relatively warm environment (Arkell, 1935). For more recent reefs, such as those dated Neogene, indications of a tropical environment may be sounder, as in reefs containing a coralline fauna well represented in the Mediterranean area during the Lower Miocene, decreasing in the Upper Miocene, and disappearing before the beginning of the Pliocene (Chevalier, 1977). These indications are even more conclusive for the Holocene: Hamada (1977) described the occurrence of tectonically raised reefs around Tokyo Bay, Japan, dated at 4200–6900 BP, and as hermatypic corals no longer live in Tokyo Bay, it is deduced that these reefs are evidence for a deterioration of the climate since a Holocene optimum. The more recent the facts, the more convincing the evidence.

CHAPTER TWO

Surface features of coral reefs

In this chapter, the surface features of coral reefs will be described along a cross-section from the outer slope to the inner side. This description will include the most common features; the peculiarities producing the different reef types will be indicated in Chapter four. Some general problems, such as those related to the formation of beach rock and the influence of hurricanes on reef morphology, will also be discussed here.

1. GENERALIZED CROSS-SECTION

1.1. Outer slope

The outer slope of the reefs is, as a rule, very steep, although the gradient differs according to sites. The differences concern principally the length of the slope, which varies in relation to the depth of the sea-floor upon which the reefs have been built. Some frameworks rise from a sandy sea floor where the depth does not exceed a few metres, but even there the reef slope is steep: the corals rise abruptly as living colonies forming a kind of staircase, as, for example, off Noumea in the New Caledonia lagoon, where reefs are built on a sandy floor 9 to 22 metres deep (Faure *et al*., 1981).

More significant, however, are the slopes surrounding oceanic reefs, which rise several thousands of metres from the deep sea-floor. At Bikini Atoll (Emery *et al*., 1954), the outer slope descends to 3600 or 4500 metres, the maximum and average slopes being 46° and 36° respectively. On Rongelap, Rongerik, and Ailinginae Atolls, also in the Marshall Islands, the same

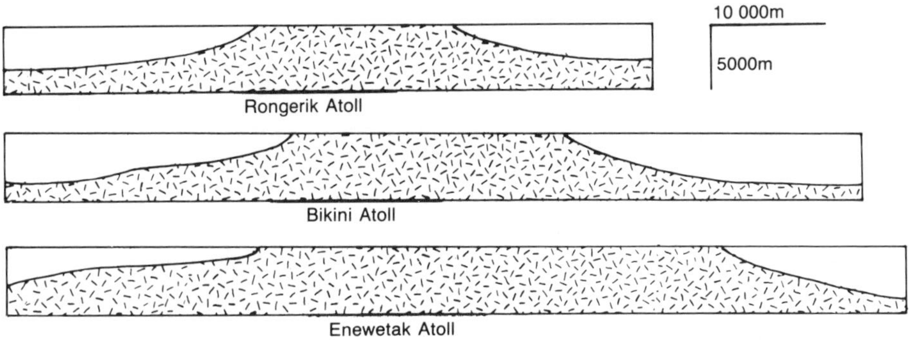

Fig. 6 Cross-profiles of three atolls in the Marshall Islands, according to Emery, Tracey and Ladd (1954). Heights and lengths at the same scale

authors reported average slopes of 31°, 31°, and 35° (Fig. 6). The leeward slopes at Bikini are steeper (37°) than those to windward (33°). At Mururoa Atoll in the Tuamotu Islands (Chevalier *et al.*, 1968), the slope averages 45° from 20 metres to 300–500 metres depth, and then decreases to 30°–35° in the north and 15–25° in the south, the relation with wind direction being less evident than at Bikini. On the barrier reef of the Gambier Islands, Eastern

Fig. 7 A steep slope with overhangs and a school of fish in the upper edge of a Caribbean reef (Photo R. Murphy, Cousteau Foundation)

(a)

Fig. 8(a) *Acropora palmata*, typical of uppermost reef slope in the Caribbean, at Sainte Luce reef, Martinique. Depth: 3 metres (Photo J. Laborel; (b) *Montastrea annularis*, with a subsidiary growth of *Porites porites*, at a depth of 6 metres on Sainte Luce reef, Martinique (Photo J. Laborel)

Polynesia (Brousse *et al.*, 1974), the slopes, exceptionally gentle as far down as 30 metres, are between 12° and 40° from 30 to 100 metres and between 16° and 32° from 100 to 1000 metres.

In many cases, terraces interrupt the slope in its uppermost section. A terrace is of general occurrence in the Marshall Islands (Tracey *et al.*, 1948; Emery *et al.*, 1954), where it occurs between 15 and 21 metres, and has been referred to as the 10 fathom terrace. Its greatest width is 3.6 kilometres, and it contains shallow irregular depressions. Terraces also exist at Mururoa in the Tuamotu Islands, at 8 and 20 metres depth, with some irregularities (Chevalier *et al.*, 1968), and they have also been recognized at the same levels at Raroia in the same archipelago (Newell, 1956). Thus the precipitous slopes may not start from the reef edge. At Johnston Island in the Central Pacific, terraces or platforms are equally reported at 9 and 18 metres depth, with numerous ridges or knolls, and many sink holes in the upper one (Keating, 1985). Such terraces, however, are not found on all reef slopes. In the Red Sea they do not occur on the reefs of the northern part of Farsan Bank (Guilcher, 1955), where they have not been found at Shab Suleim, Malathu and Marmar. At Shab Suleim, at least in the upper part, the eastern, inner side, is steeper than the western, windward side, as at Bikini. But on all these Red Sea reefs the slopes average 45–50° down to 200–500 metres.

A feature common to the upper few metres of outer slopes is the occurrence of many overhangs and caverns, inhabited by brilliant and numerous schools of fish (Fig. 7), the overhangs testifying to the active growth of coral outwards. The most beautiful pictures taken by divers to portray coral reefs are from this part of the hermatypic frameworks. The caverns are often connected with the inner part of grooves which they extend landwards under the reef surface. On Caribbean reefs as in Cuba, Bonaire, St Croix, Grand Cayman and Martinique, *Acropora palmata* (Fig. 8a) has been able to build palmed or branchy ramifications near the surface, which can withstand heavy surf (Kühlmann, 1974; Scatterday, 1974; Adey, 1975; Roberts, 1974; Laborel and Bouchon, 1986). Also in the West Indies, growths of *Acropora cervicornis* occur just below at ±5 metres, followed by *Montastrea annularis* (Fig. 8b) at 6–10 metres, both at depths where wave attack is somewhat reduced. In other areas, similar vertical zones of coral species are also found. Investigations of the windward side of oceanic reefs are generally very difficult because of the surf; nevertheless, where divers have explored these environments, as on Aldabra Atoll in the Indian Ocean (Drew, 1977), they have confirmed a very marked decrease in branching corals and Alcyonariae and a sharp increase of foliaceous and encrusting types below a depth of 30 metres.

1.2 Outer edge: spur and groove systems and algal (*Porolithon*) ridge

The outer edge of the reefs on the windward side, from the surface down to a few metres, is generally cut by systems of more or less spaced grooves, separated by spurs. The classic description is from Bikini Atoll, where 'the seaward slopes are cut by well-developed grooves normal to the reef front, 50 to 100 feet long, 3 to 10 feet wide, and 6 to 25 feet deep. The grooves are relatively straight but in some places are forked, usually downslope. The spurs separating the grooves are flattened algal ridges 25 to 50 feet or more in width. The tops of the ridges are covered by living algae, and the sides also appear to be algal . . . Reefs of this type are best developed on the windward side of the atoll' (Tracey *et al*., 1948; similar description in Emery *et al*., 1954). The algae which coat and partly form the spurs and grooves are pink in colour and belong to the genus *Porolithon*, which includes *P. onkodes*, smooth or slightly undulating, and living especially in heavy surf, and *P. craspedium*, with an aspect of cauliflower, thriving mostly in more sheltered corners or caverns. The upper part of the system forms a ridge or broad arch, commonly from 5 to 15 metres wide, and rising some 0.3 to 0.6 metre above the lowest part of the backing reef flat. This *Porolithon* ridge is constantly beaten and washed by the surf, which is necessary for *Porolithon* to live.

This *Porolithon* ridge, shaped in spurs and grooves, is found in the other Marshall Islands; it is not found on Tarawa and Abemama at Kiribati (Guilcher, 1967b), perhaps because these atolls lie in the doldrums so that the surf is moderate (but spurs and grooves occur there too: Figs. 9 and 10). This feature appears on many other Pacific archipelagos, especially at Funafuti,

Fig. 9 Northeastern corner of Tarawa Atoll, Kiribati. Spurs and grooves. Pond in reef flat (Photo A. Guilcher)

Fig. 10 Tarawa Atoll, Kiribati, northeastern corner. Spurs and grooves. Fishing weirs on reef flat. Beach rock in the beach (Photo A. Guilcher)

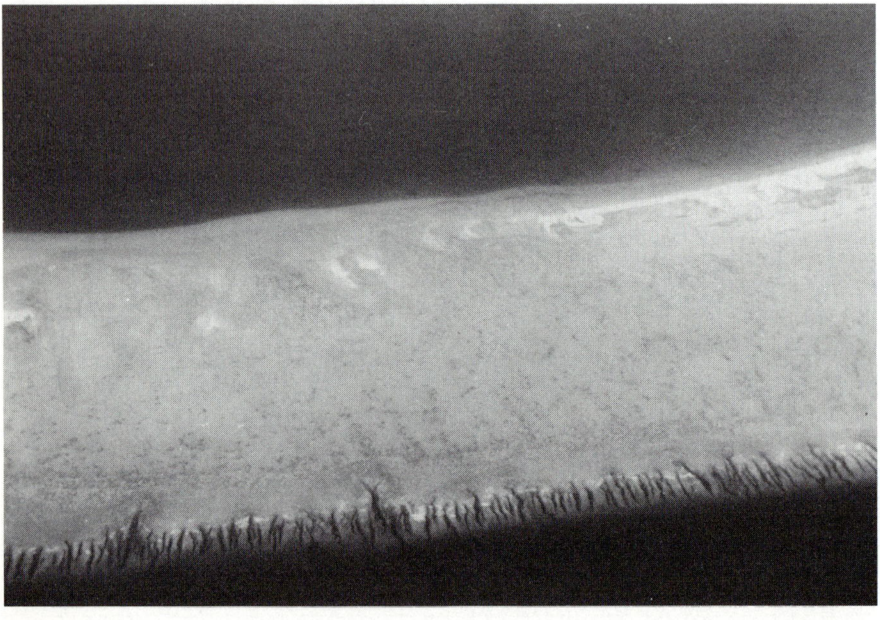

Fig. 11 Scilly Atoll, Society Islands. northeastern rim. Outer spurs and grooves in *Porolithon*; reef flat; sandy accumulations near inner edge, adapted to overwash flowing to open sea when lagoon is overfed by southern swell. See Fig. 120 (Photo A. Guilcher)

Fig. 12 Tongatabu, Tonga Islands. Terraced rimmed pools built around a blow hole on outer edge of reef (Photo F. Doumenge)

Fig. 13 Tongatabu, same place as Fig. 12. Blow hole in action in the middle of terraced pools (Photo F. Doumenge)

Tuvalu (Armstrong *et al.*, 1904); at Tutuila, Samoa; in the Society Islands at Mopelia and Scilly Atolls and at Bora Bora barrier (Guilcher *et al.*, 1969) (Fig. 11); on many Tuamotu atolls as Mataiva, Rangiroa, and Mururoa (e.g. Chevalier *et al.*, 1968; Guilcher *et al.*, 1969; Stoddart, 1969b; Delesalle, 1985). In these islands the *Porolithon* ridge rests generally upon a more or less eroded older reef. The windward reef at Kure Atoll, Hawaii, is also made of red–brown living algae with a spur and groove system (Ladd *et al.*, 1970, frontispiece).

As the spurs are prograded seawards, the intervening grooves are progressively coated by *Porolithon* in their inner parts, so that they tend to become surge channels, in which the water pushed by the uprush of the swash circulates in caverns and is compressed through blow-holes in the inner part of the ridge; this is the room-and-pillar structure described by American authors (Tracey *et al.*, 1948; Emery *et al.*, 1954), which is well represented at Rangiroa, Mopelia, Bora Bora, and others. There are variations from place to place in the width of the grooves and the height of the ridges, but they are only details.

A curious feature in several places in Indonesia and at Tongatapu, Tonga Islands (Kuenen, 1950, p. 431–432; Doumenge, personal communication) is that, around blow holes fed by surge channels, calcareous algae have built

Fig. 14 Mopelia Atoll, Society Islands. Southwestern part. Rimmed pools with edges built by *Porolithon*, and small jagged remnants of former (probably Holocene) reef. See Fig. 106 for a dated outcrop of former reef at Mopelia (Photo A. Guilcher)

'shallow, rimmed pools resembling the terrace formations around geysers and hot springs; the pools are circular and arranged in successively lower terraces as sections of a circle around the overflow from the higher levels' (Kuenen, 1950) (Figs. 12 and 13). The low rims are built by algae, but resemble the erosional rims found as residuals on calcareous outcrops around similar pools in Morocco and Madagascar, also in the intertidal zone. It is an interesting convergence phenomenon. Incipient rimmed pools, also with edges built by *Porolithon*, exist on the southern part of the reef edge at Mataiva Atoll, Tuamotu Islands, and at Mopelia Atoll, Society Islands (Fig. 14).

On St Croix, Guadeloupe, and elsewhere in the Caribbean, algal ridges are even more important outer parts of reefs than in the Pacific (Adey, 1975; Adey and Burke, 1976; Connor, 1977; Battistini and Hinschberger, 1985). They often form only a veneer over or beside a fossil Holocene or Pleistocene reef edge, a structure especially evident on Mopelia Atoll in the Society Islands, and at Mataiva, Rangiroa, Kaukura, Hikueru and other atolls in the Tuamotu Islands (Guilcher *et al.*, 1969; Delesalle, 1985).

In the Pacific, living hermatypic corals are comparatively scarce on the upper reef edges, but they generally thrive in front of and just below the spurs, as, for example, on Mopelia Atoll (Guilcher *et al.*, 1969), where the outer slope is covered by living colonies down to at least 25 metres, and at Mataiva Atoll (Delesalle, 1985) where *Acropora*, *Porites*, *Montastrea*, *Pocillopora* and other corals form many small colonies accompanied by corallinaceae and brown algae on the reef front.

In some cases, as at Taiaro Atoll, Tuamotus (Chevalier *et al.*, 1976), the *Porolithon* ridge is not well developed on sheltered sectors, where scleractinian corals are abundant in the spur and groove system, but it reappears in areas exposed to strong surf.

Outside the Central and South Pacific, spurs and grooves have been reported from many reefs. They are well represented on the Mayotte barrier reef in the Western Indian Ocean (Guilcher *et al.*, 1965), where the length of the spurs varies from 40 to 80 metres, and the intervening distance from 12 to 24 metres. In places where the barrier reef is now drowned under several metres of water, the length increases to 200–900 metres and the spacing to 23–50 metres. Well-developed spurs and grooves are also present on Farquhar Atoll (Battistini *et al.*, 1979), on Grand Récif off Tulear on the southwestern coast of Madagascar (Gravier *et al.*, 1970; Pichon, 1972, 1974; Weydert, 1974). At Shab Suleim on the Farsan Bank in the Red Sea (Guilcher, 1955), they occur on the western, windward side of the reef, and are absent from the leeward side, much as at Bikini. Off Tulear, Weydert showed how the grooves, which extend down to 18 metres, display an intersecting pattern in their middle sectors. But, if algal ridges do exist on some reefs in the Indian Ocean, as at Diego Garcia Atoll and on the Maldives (Stoddart *et al.*, 1971; Davies *et al.*, 1971), they are generally missing from reefs lying outside of the northeastern part of that marine area, at Mayotte, Madagascar, Mauritius, and are poorly developed in the Seychelles (Guilcher,

1971, p. 75–76; Braithwaite, 1971). A consequence is that spurs and grooves on reefs in the Western Indian Ocean and on Farsan Bank are neither built of, nor coated with, *Porolithon*, but are of large hermatypic corals. The bottom of the grooves is often, at least partly, covered with coralline sand.

Spurs and grooves have also been described from reefs in the West Indies. Stoddart (1962) reported them from Belize in Central America, again on the windward side, where the spurs were made of vigorous branching corals, chiefly *Acropora palmata*, with no indication of algal blanketing. Roberts (1974) described them around Grand Cayman Island, south of Cuba, as systems associated with two fore-reef terraces found respectively at 5 metres and 15 metres depth above a steeper slope. As elsewhere, they characterize the reefs on coasts exposed to high wave energy, and are poorly developed on the more sheltered west coasts. There is an angular disparity between the spurs and grooves on these two terraces as a result of adjustment to variations in wave refraction. At Tulear, too, where the southwesterly swell is oblique to the reef edge, the orientation of the spurs is a resultant between the alignment of the reef edge and the direction of the swell (Weydert, 1979).

So far, we have mentioned only spurs and grooves on the edges of reefs rising from great or moderate depths, but similar features have also been described from the edges of reefs rising from shallow sandy areas. Such is the case along the northwestern coast of Madagascar, for example at Nosy Foty (Guilcher, 1956), where the depth in front of the reef does not exceed 14–16 metres, and at Nosy Valiha, Radama Islands, where the nearshore depth is similar (Guilcher *et al.*, 1958, p. 127). A close examination of the reef at Nosy Valiha has shown the same pattern as at Tulear, on the Farsan Bank, and elsewhere, with vigorous *Acropora* colonies covering the spurs, and sand and coral debris flooring the grooves, which are 5 to 6 metres deeper.

At Bikini Atoll, Munk and Sargent (1948) explained the spur and groove systems as an adaptation of the reef front to the surf: the spurs disperse the energy of the waves and allow the organisms to grow, utilizing oxygen and nutrients from the constantly stirred water. Newell (1954), describing the system at Raroia, Tuamotus, wrote that it ‘clearly combines both erosional and depositional processes, and it is probable that it is also true of reefs in general’: as growth occurs on spurs, ‘scour by sand is the chief agency of erosion along the floors of the grooves . . . that are gradually extended headward, in many places reaching well onto the reef flat as very shallow, more or less bare, furrows, embryonic surge channels’. This statement, made for an oceanic reef, is equally true for a coral reef in shallow water, as at Nosy Valiha in Madagascar, where the same relation between growth of spurs or buttresses, and erosion of grooves or furrows, appears valid (Guilcher *et al.*, 1958). At Grand Cayman Island, the grooves ‘provide avenues through which sediments produced in shallow water environments can be transported to deep off-shelf sites of deposition’ (Roberts, 1983), and this has also been observed at Tulear (Gravier *et al.*, 1970, Weydert, 1974).

1.3. Detrital outer ridge

Many fore-reefs bear on their outer edge detrital accumulations instead of, or beside, growing coralline or algal structures. Such accumulations stand slightly (one metre or less) above the inner reef flat which will be described below (Figs. 15 and 16). Both types of ridges are found off northwest Madagascar, a detrital ridge existing, for example, on reefs rising from shallow depths (around 10 metres) at the entrance to Baie du Courrier (Guilcher, 1956) (Fig. 17). Huge blocks, which Matthew Flinders and others called negro-heads or nigger-heads, are common, eroded from the living reef and thrown on to the reef flat during tropical hurricanes. Such blocks can even be pushed as far as the inner side of coral reefs by exceptionally strong cyclones (Fig. 31). At Grand Récif near Tulear (Weydert, 1973; Pichon, 1972, 1974; Battistini *et al*., 1975), which receives a heavy swell originating in the Southern Ocean, blocks form either ramparts or series of elongated domes perpendicular to the general direction of the reef front, along which the grain size decreases toward the inner ends (Fig. 20). Some authors, quoted in Battistini *et al*. (1975), have called them 'crags and tails', by analogy with the shape of glaciated structures in Scotland, but the 'crags' here are not made of *in situ* rocks, but of big blocks. Detrital outer ridges are also common on the smaller reefs south of Tulear, but at Itampolo their place is exceptionally taken by a small algal ridge (Battistini, 1964). Many subtypes

Fig. 15 Mayotte barrier reef, western Indian Ocean. Low spring tide. Detrital outer ridge with coarse material and some living colonies showing that heavy surf is infrequent (Photo A. Guilcher)

Fig. 16 Abemama Atoll, Kiribati. Detrital outer ridge (Photo A. Guilcher)

Fig. 17 Patch reef off Baie du Courrier, northwest coast of Madagascar (location on Fig. 80). Large blocks on outer, northwestern edge of reef (scale given by scuba on right), moved by infrequent hurricanes, between which oyster banks can grow on tops. Low spring tide (Photo A. Guilcher)

exist, and sometimes the ridge is flat, consisting of sparse blocks of various sizes scattered over the outer reef, as at Mayotte (Guilcher *et al*., 1965a).

Just behind the outer ridge, and distinctly lower, there is frequently a subzone of living corals, often very vigorous, alternating with encrusting algae. Two conditions are favourable for the coral colonies: they are sheltered from the heaviest surf, although the water is still well stirred and oxygenated, and they emerge only at the lowest spring tides.

1.4. Reef flat

The reef flat generally forms the widest part of a coral reef. Its width can vary from less than 100 metres to several kilometres. An example of wide reef flat is found on the fringing reef of Rodriguez Island in the Indian Ocean, which reaches 2000 metres in width in several places (Faure, 1974; Montaggioni, 1974). As a rule, the reef flat consists mainly of a pavement cemented by calcareous algae and incorporating various detrital elements derived from the coral community, such as coral fragments, articles of Halimeda, and foraminifera. The reef flat is generally exposed to the atmosphere at low spring tides, but there are many pools of various sizes, often with overhanging sides, some decimetres deep, which provide habitats for soft algae, gastropods, lamellibranches, echinoderms, starfishes, and small colonies of living coral. Loose elements such as sand or gravel formed from the branches of broken *Acropora* partly floor these pools (Figs. 18 and 19).

Fig. 18 Inner side of Mayotte barrier reef, Indian Ocean. Large *Halimeda* meadow, a major source for sedimentation on many reefs (Photo A. Guilcher)

Fig. 19 Current aspect of reef flat on the fringing reef near Mombasa, Kenya. Detrital coral branches cemented by calcareous algae. Shallow ponds with overhanging rims, in which small colonies, gastropods and starfish live (Photo A. Guilcher)

The most interesting feature of reef flats is their *striation*. They currently bear series of long stripes, consisting alternately of consolidated reef flat and of sand and gravel. These stripes have been described and illustrated from Tulear Grand Récif (Caye and Thomassin, 1967; Weydert, 1973, 1974; Pichon, 1972, 1974; Battistini *et al*., 1975), smaller reefs in northwest and southwest Madagascar (Guilcher, 1956; Battistini, 1964), and Mayotte barrier reef (Guilcher *et al*., 1965a) (Figs. 20 and 21). They run at right angles to the general trend of the reef and appear quite distinctly on air photographs as white and dark bands, the white being sand and gravel. At Mayotte they clearly continue across the reef in the alignments of the grooves at the outer edge, and the difference in height with the solid stripes varies from 10 to 50 centimetres. As with the grooves, the sand and gravel stripes are controlled by wave dynamics: where the outer edge of the Mayotte reef shows re-entrants, and refracted waves arrive in a modified pattern, the direction of the stripes shifts accordingly. At Tulear (Pichon, 1974) 'there is a balance, on the reef flat, between the total area covered by reef builders as a whole, and the total area covered by sedimentary deposits. All these deposits are of skeletal origin, and detailed surveys at regular intervals over a total period of twelve years do not show any noticeable change in their position and extent on the reef flat', so that a state of equilibrium with hydrodynamic factors has apparently been reached.

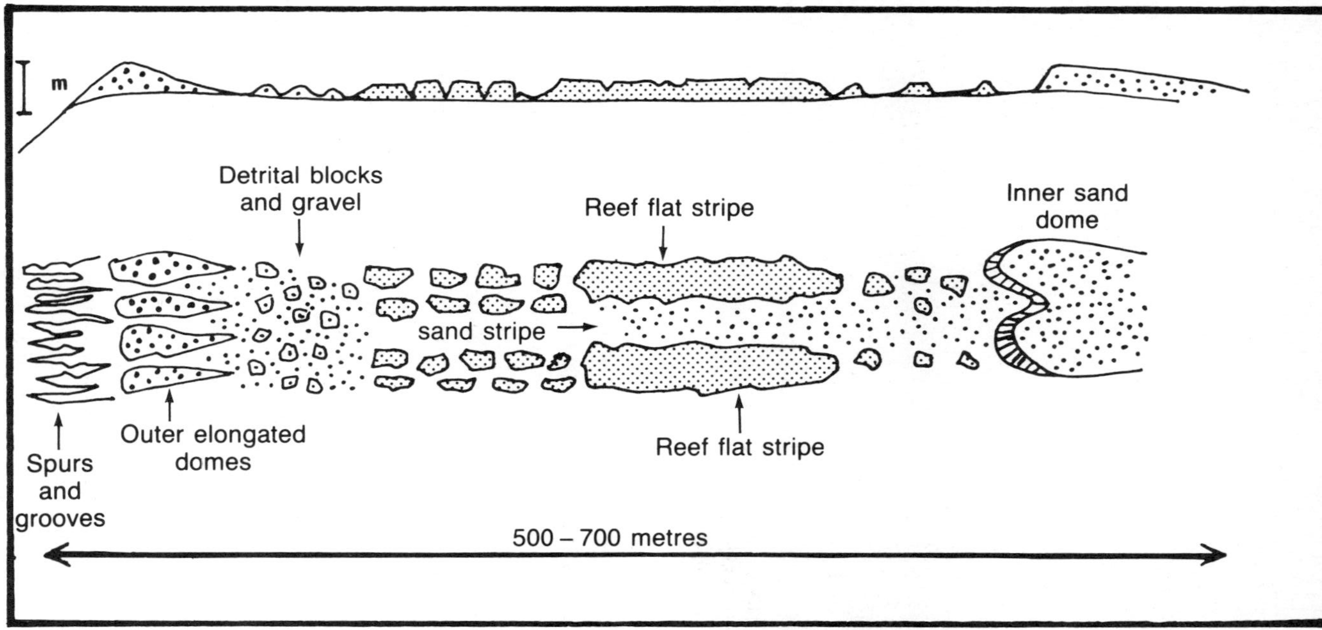

Fig. 20 Reef pattern (section and plan) across the central part of Toliara (Tulear) barrier reef, southwest Madagascar, according to Weydert (1973)

Fig. 21 Mayotte barrier reef, Indian Ocean, southern part. Outer spurs and grooves, stripes (striations) on the reef flat, coral knolls on lagoonside (Photo A. Guilcher)

In spite of the fact that the reef flat has generally a very hard surface, resembling a pavement, the reef in its internal parts is a looser mass, extremely porous to water, with uncemented areas alternating with more lithified lenses in an irregular pattern. Such heterogeneous structures have been accurately described in atolls where deep drillings have been made as at Enewetak in the Marshall Islands (Oberdorfer and Buddemeier, 1985) and Mururoa in the Tuamotu Islands (Aissaoui and Purser, 1985). Some of the consequences will appear in Chapter four, section 4.7.

Although variable thoughout the coral world, the width of the reef flat is more or less constant on atolls in a number of areas. Tayama (1952, p. 250) found it to average 600 metres in the Palaus, 500 metres in the Carolines, 500 to 530 metres in the Marshalls. The reef flat tends to be wider where the lagoon is shallow, narrower where it is deep.

1.5. Microatolls

A special feature that has received much attention is the existence of microatolls, not exclusively but frequently found on reef flats, particularly in pools where water is retained at low tide. An excellent review of the morphology and terminology of microatolls has been made by Stoddart and Scoffin (1979). As other authors, especially Scheer (1971), pointed out, the term microatoll should strictly be reserved for individual subcircular coral colonies, not exceeding 6 metres in width, and often much smaller; it should not be used for the larger corallian structures, also subcircular, for which the terms faro and atoll are more appropriate. Unfortunately such a restriction has not always been observed, so that there is some confusion in the literature.

'Typical microatolls comprise single colonies of massive corals, especially of *Porites*, usually round (though complexity of form is stressed by Kuenen, 1933, 64) and with a flat or concave upper surface devoid of living polyps' (Stoddart and Scoffin, 1979). There are frequently several merging circles, instead of only one. Other scleractinians, such as Acroporae, can also form microatolls. The alcyonarian *Heliopora* and algae may grow sporadically over the structure, or even play an essential role in shaping it, as on the Great Barrier Reef.

Several explanations have been proposed for microatolls. The first one is a sedimentation heavier than usually, leading to a differential growth: fine particles and suspension in sea water settle on the top of the colonies and kill the central polyps, which continue to grow on the outer vertical surfaces where the ciliae prevent deposition (Fig. 22). This explanation was put forward, for example, by Wood-Jones (1912) for Cocos–Keeling Atoll in the Indian Ocean, and by Krempf (1927) for frameworks off the Vietnamese

Fig. 22 Nosy Langna near Analalava, northwestern coast of Madagascar. Microatolls living in turbid waters off the mouth of River Loza. Low spring tide (Photo A. Guilcher)

coast. It is also valid at Mayotte (Guilcher *et al*., 1965) as a result of fine sediment derived from erosion of lateritic soils on the adjacent mainland.

A second explanation is that colonies are limited in upward growth by the level of low tides, so that lateral growth prevails, and the central area dies. Kuenen (1933) and Umbgrove (1947) give both explanations for Indonesian reefs. A third explanation, suggested for alcyonarian atolls in American Samoa (Cary, 1931), is that nutrient supplies are greater round the periphery than in the centre. Dead microatolls standing at abnormally high levels may have grown in former pools which have been drained, or they may have formed during a phase of higher relative sea-level.

Microatolls are especially frequent on the Great Barrier Reef, where their origin(s) were discussed by Hopley (1982, p. 109–115). In the coral world as a whole, the present writer thinks that the first explanation is generally valid.

1.6. Back reef; boat channel

The area behind the reef flat differs on fringing reefs, barrier reefs, and atolls. In the second and third type, is is replaced by a lagoon, usually several tens of metres deep, which will be discussed in Chapter 4, section 4.3. Behind fringing reefs, there is frequently, but not always, a depression, up to a few metres deep at low tide, between the reef flat and the mainland shore, which

Fig. 23 Fringing reef in Kenya near Mombasa. Crescentic pools in inner part of boat channel, with Phanerogams. Low spring tide (Photo A. Guilcher)

has only poor coral growth, probably because of deposition of terrigenous sediment from the continent, or the island on which the reef is anchored. This depression, which was mentioned and explained in this way by Darwin (1842, chapter three), is called the boat channel because it allows a coastal traffic with small boats: an example is found near Mombasa in Kenya (Bird and Guilcher, 1982). A boat channel existed just north of Jeddah in Saudi Arabia (Guilcher, 1985a) before it was artificially infilled during the seaward extension of that city. In the southern and southwestern parts of Rodriguez, Indian Ocean, the boat channel can exceed 5000 metres in width in places (Montaggioni, 1974). Boat channels also exist in Madagascar (Guilcher, 1956): see Chapter four, sections 1.2 and 1.3, for details.

Parts of these shallow depressions are colonized by meadows of marine phanerogams, which are also sometimes found on the inner parts of reef flats and lagoons connected with barrier reefs, as at Tulear where *Cymodocea* and *Thalassia* spp. form prairies (Weydert, 1968; Pichon, 1974) and at Mayotte (Guilcher *et al*., 1965). In northwest Madagascar and near Mombasa in Kenya, these meadows are partly associated with sets of crescentic pools (Fig. 23), the phanerogams growing on the sides of the ridges between the pools. Such patterns result from wave action at high tide, when the ridges and pools are submerged. They seem to require a tidal range of at least 2 or 3 metres at spring tides. They have not yet been reported from areas outside the Western Indian Ocean.

In some places, the back reefs bear extensive meadows of *Halimeda*, whose small discoid plates, released after death, provide an important part of the carbonate sedimentation. This is especially true in the northern region of the Great Barrier Reef of Australia (Drew and Abel, 1985; Orme, 1985), where these meadows cover at least 2000 km^2 (see also Fig. 18). *Halimeda* platelets are frequently extremely important in the sedimentation of other parts of reefs, for example in Brazil (Laborel, 1969) where they accumulate in great numbers. *Halimeda* is extensive on the inner side of the barrier reef at Guadeloupe Island in the Caribbean, at depths of up to 3–4 metres (Guilcher and Marec, 1978), and on the deep fore-reef of Enewetak Atoll in the Marshall Islands, where it is a major carbonate producer, as it is also on the fringing reefs of Jamaica (Hillis-Colinvaux, 1974, 1985). In other places, algae belonging to the genus *Caulerpa* have a wide distribution, for example at Takapoto Atoll in the Tuamotu Islands, where they cover the outer slope all round the atoll from 40 metres to more than 70 metres depth (Meinesz *et al*., 1981).

On the coastline behind a fringing reef, in the innermost parts of the boat channels where the surf is sufficiently weakened, mangroves are able to grow, usually belonging to the genera *Avicennia* or *Rhizophora* (e.g. in the Gulf of Elat or Aqaba) (Guilcher, 1979a) (Figs. 73, 74, and 77), but including sometimes *Laguncularia* and *Conocarpus*, as in Paraiba on the northeastern coast of Brazil.

2. SANDY ISLANDS: VEGETATION

Several types of sandy islands exist on reefs, or in association with them. They will be described in detail in Chapter 4, but it is necessary to characterize their vegetation cover before the effects of hurricanes are considered.

This vegetation has been studied in great detail, especially by Fosberg and Sachet (e.g. Sachet *et al.*, 1983; also numbers 274 to 277, 1983, of Atoll Research Bulletin, equally by Sachet; for the West Indies, references to Stoddart, 1962a, p. 112; for the Indian Ocean, Stoddart *et al.*, 1979; general recent review with many references by Lee, 1984).

Summarizing the facts, the natural vegetation is usually quite dense except just behind the shoreline. In the Society and Tuamotu atolls for example (Figs. 24 and 25), where the islands are directly exposed to surf and spray, the outermost fringe of plants consists of a salt- and drought-resistant *Pemphis* scrub, with dwarf bushes scattered on rock surfaces, becoming taller and much thicker just behind, where they are associated with *Scaevola* and *Suriana*. Farther inland was originally an inextricable broad-leaved, mesophytic forest, several metres high, including *Tournefortia*, *Pandanus*, *Pisonia*, *Guettaria*, and other species. This forest still exists in places, but generally it

Fig. 24 Mopelia Atoll, Society Islands. Eastern, exposed side of atoll rim. Dwarf *Pemphis* on coral conglomerate in foreground, passing quickly into a thicker and higher vegetation in background (*Tournefortia* and cultivated coconut trees). Picture taken in 1963, when there had been no hurricane for several decades (Photo A. Guilcher)

Fig. 25 Rangiroa Atoll, Tuamotu Islands. Landscape of a cultivated atoll (Photo A. Guilcher)

has been replaced by coconut plantations, in which the undergrowth deriving from early stages in the revival of the former vegetation is periodically cleared. On similar islands in the Indian Ocean and in the West Indies, the natural vegetation was the same broad-leaved forest, but there too coconuts have been widely planted. Although the climate is very dry, the islets scattered on the Farsan Bank in the Red Sea are covered by a thick spiny growth, more than 2 metres high. In more sheltered positions, such as on islands situated in lagoons, the natural vegetation is extremely dense just behind the beach.

3. BEACH ROCK

Beach rock is commonly found on sandy cays and in beaches behind coral reefs (Figs. 26, 27, and 28). According to a definition adopted in a recent symposium (Dalongeville, 1984, p. 161), beach rock, which occurs on beaches as a series of flagstones sloping seawards, is a sedimentary formation indurated at the mediolittoral level by a calcareous cement, consisting initially of aragonite or calcite. There are fossil beach rocks, lying now at higher or lower levels. Beach rock does not exist only in association with coral islands: it occurs in many calcareous beaches around the Mediterranean, and perhaps at even higher latitudes. Numerous references on beach rock are included in

Fig. 26 Mataiva Atoll, Tuamotu Islands. Massive and multiple slabs of beach rock on western side of the rim, apparently belonging to the same generation (Photo A. Guilcher)

Fig. 27 Nosy Foty, a sand cay off the northwestern coast of Madagascar. Two sets of beach rock with different dips (Photo A. Guilcher)

Fig. 28 Mataiva Atoll, Tuamotu Islands. East coast. Three superposed, successive beach rocks: the older one, much dolomitized, in the foreground, and two younger ones on the right and in the middle distance (Photo A. Guilcher)

Dalongeville, 1984, p. 169–197; see particularly Scholten, 1972; Russell, 1962; Guilcher, 1961; Stoddart and Cann, 1965. The origin of beach rock remains largely a matter of discussion. Russell (1962) said that cementation consisted of the deposition of calcite, but in many exposures, especially in coral areas, aragonite has been found to be the primary cement, and in other places both occur (Stoddart and Cann, 1965). Dalongeville and Sanlaville (in Dalongeville, 1984) stated that, around the Mediterranean, beach rock was exposed on retreating shores, and is still forming within unconsolidated Holocene deposits which have appeared on the surface as a result of coastal retreat; such a view has seemed (Guilcher, in Dalongeville, 1984) partly true but too exclusive, especially for tropical areas where coral reefs thrive. The action of ground water has been invoked, but beach rock is present in sites where there is no ground water table because of aridity, or for topographic reasons. A debated problem is the relative age of inner and outer slabs: at Aqaba in the Red Sea, the most recent slabs occur on the outer side, and at Tulear, Madagascar, the upper ones are the most recent; but on the Great Bahama Bank the inner ones appear to have been cemented subsequently; at Hao in the Tuamotus, the inner slabs are less consolidated, but may not necessarily be younger. It is not yet clear whether the flagstone pattern existed as beach layering prior to the consolidation, or whether it develops during the hardening (Dalongeville, 1984, p. 163–165).

Where it is exposed on the shore, beach rock is currently being undercut by wave erosion, and as this washes out the underlying loose sediment, the layers are undermined, and fractured into slabs which can either slump to subtidal levels, where beach rock does not form, or be thrown by storm waves or tsunamis to supratidal levels and piled up as a mass of broken rock. Evidence for former sea-levels can only be derived from *in situ* beach rock and not from isolated fragments, even if these are very large. On the other hand, disintegrated fragments of beach rock are frequently incorporated into new beach rock layers (Strasser and Davaud, 1985). In a seminar held in Tahiti (Davies and Montaggioni, 1985), some participants have warned against the use of beach rock to determine former sea-levels, while others thought it may be a good indicator, especially in areas of low tidal range.

Undoubtedly there are places where beach rock is still being formed, but beach rocks have been recorded from very remote times, perhaps as early as the Proterozoic (Donaldson and Ricketts, 1979). Beach rock patterns are useful in the study of variations of sandy beach outlines, especially sand cays on coral reefs, since they indicate former shorelines, and can be used to determine the trend of changes in progress, as shown by Battistini and Guilcher in Dalongeville (1984, p. 11–18 and 67–76); Stoddart (1975), from Tongatapu; and Stoddart (1969e), from Guadalcanal in the Solomons. In northeast Brazil, parallel outcrops of Holocene beach rock, now slightly below sea-level, have formed basements of coral reefs which consequently occur in parallel rows. These unusual Brazilian structures, already described by Darwin at the end of the second edition of his book on coral reefs, are locally known as *arrecifes*, a Portuguese word which thus designates sandstones rather than true coral reefs, although the latter are connected with the former (Laborel, 1969; Dalongeville, 1984, p. 77–83). See also Chapter four, section 7.

Beach rock is associated with sandy deposits on coral reefs, but can also incorporate pebbles, sometimes large, and not exclusively coralline: pebbles of volcanic origin have been supplied to coral beaches in Madagascar and cemented into beach conglomerates (Guilcher, 1961).

Beach rock should not be confused with what has been called *cay sandstone* by Kuenen (1933, p. 87–88, 1950, p. 433–435) in the East Indies. Both can occur on coral reef islands, as on Belize in the Caribbean (Stoddart, 1963, p. 108) and at Diego Garcia Atoll in the Indian Ocean (Stoddart and Taylor, 1971, p. 10 and 19), but cay sandstone consists of horizontal strata, while beach rock is characterized by seaward dipping layers. Like beach rock, cay sandstone occurs outside the regions of coral reefs, for example in the sandy Tiger Peninsula (Restinga de Los Tigres) in South Angola (Guilcher *et al.*, 1974), where coral growth is impossible because of the Benguela cold current. It seems to result from vertical percolation of freshwater through the beach sand during heavy occasional rains, causing solution and then deposition of carbonates.

4. CALCIUM CARBONATE BUDGETS

Such budgets have been established for particular reefs. Discussion and results of studies of annual energy budgets and calcification mechanisms may be found in Gladfelter (1985). An example was based on a study of a fringing reef on the west coast of Barbados in the Caribbean by Scoffin *et al*. (1980). On this reef, measuring 7 hectares, the overall balance is favourable, with a calcium carbonate production of 206 tons/year, against a destruction of 123 tons/year. In this case algae, including *Halimeda*, play only a minor part in carbonate production, while elsewhere they have a more important role. Corals provide more than one half of the carbonate sands (43 to 93 per cent), followed by Corallinaceae, cryptocrystalline particles, molluscs, foraminifera, and echinoderms. Terrigenous elements on this reef never exceeded 3.3 per cent, but they are more abundant on many fringing reefs, as in Kenya (Bird and Guilcher, 1982). Mechanical erosion is very small, since hurricanes are rare in Barbados, but again this factor is quite different on reefs in other areas (see this Chapter, section 7). Bioerosion, on the contrary, is important in that it controls the grain size, the shape and the texture of the particles. Sponges are the most efficient boring organisms, but *Cliona* spp., molluscs, sipuncules, polychaetes, and crustaceans also contribute. As a whole, erosion by boring amounts to 25 tons/year. Grazing by fishes is here quite negligible: 0.4 ton/year, but elsewhere many observers, including Darwin, have stressed this kind of erosion (see later in this section). Conversely, urchins are extremely active, especially *Diadema antillarum* which destroys 97.5 tons/year of epilithic algae.

Another example of active boring by echinoids has been noted at Enewetak in the Marshall Islands, where they accomplish bioerosion of 80 g/m^2/year on corals in the lagoon and 325 g/m^2/year on the outer reef flats (Russo, 1980). This author considers, like Scoffin *et al*. (1980), that the sponges are probably the main bioeroding agents on tropical coral reefs, but it is necessary to stress the variability of these budgets from one reef to another, and regionally.

At Lizard Island on the Great Barrier Reef, Kiene (1985) found that the rate of bioerosion was markedly different on the reef flat, where it averaged 0.41 kg/m^2/year, on the lagoon floor where it averaged 1.95 kg/m^2/year, and on the reef slope where it averaged 2.71 kg/m^2/year. There were variations from year to year, but the differences between the three sites remained: in other words, the slope is always being more rapidly eroded. Kiene found also that fishes were important in grazing, in addition to boring by molluscs, sponges and worms.

Randa (1974) has reviewed the effects of fishes on coral reefs in many different areas. Plectognath fishes (triggerfish, filefish, puffer) have powerful jaws and specialized dentition, which make them quite able to bite off corals. The polyps appear to regenerate, so that the food resource for these fish is not destroyed. Though parrotfish has been cited as feeding extensively on live coral, most studies have shown that it lives mainly on algae which it scrapes from dead coral rock, cutting into the surface as it does so. The amount of sediment produced is enormous. The surgeonfish also contributes to sedimentation by feeding in part on coralline Algae, and ingesting inorganic sediment with its algal food. Fishes that eat coral are among the most highly evolved of modern teleost fishes, a feature which suggests that this mode of feeding has developed relatively recently in geologic time. In Hawaii (Hixon and Brostoff, 1981), the grazing of coralline algae by parrotfish and surgeon fish is so intense that it changes the algal succession by allowing only certain forms to persist, while the succession in other areas follows a relatively undisturbed sequence.

Calcium carbonate production in reef environments does not decrease steadily with increasing latitude, as might be supposed. At Houtman's Abrolhos in Western Australia (29°S), average carbonate production amounts to 32 mg/m^2/day, with seasonal fluctuations, a figure not much lower than in equatorial areas (Smith, 1981).

5. BIOEROSION IN EMERGED REEFS

When reefs have emerged to form raised fringing reefs, barriers or atolls, the living reef backs against a cliff cut into the dead reef. Such situations, which will be described in Chapter three, section 7, exist in a large number of places

Fig. 29 Abulat Island, Farsan Bank, Saudi Arabia, Red Sea. Pleistocene emerged ridge reef. Two fossil corrosion notches in mushroom stacks. A folded double metre gives the scale (Photo A. Guilcher)

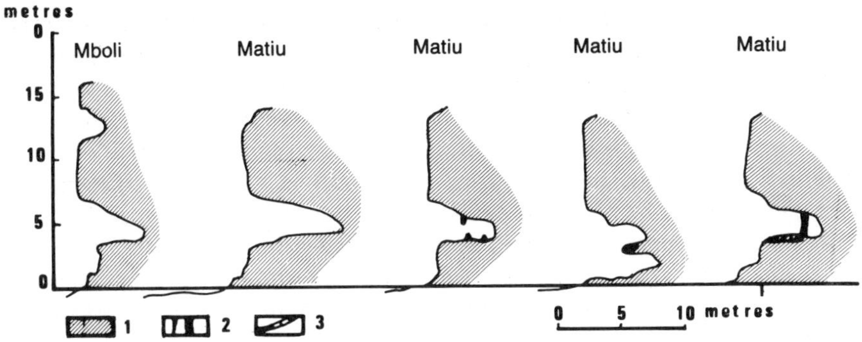

Fig. 30 Profiles with deep notches in the Marovo raised barrier reef at Mboli and Matiu Islands (located on Fig. 65). 1: coral limestone of raised barrier. 2: stalactites and stalagmites in notches. 3: algal limestone on floors of the notches. After Stoddart (1969)

including Kenya, Papua New Guinea, Vanuatu, the Solomon Islands, and the Farsan Bank in Saudi Arabia (Bird and Guilcher, 1982; Guilcher, 1955, 1974; Chappell, 1974a; Stoddart, 1969c,d). In these cases, it is common to find the old coral reef cut into by cliffs with deep notches, sometimes with more than one notch in the same cliff, with spectacular overhangs or visors (Figs. 29 and 30). The surface of the rock is always extremely jagged, as are also the residual, sharp pinnacles (the so-called feos) of old corals, as seen on Rangiroa and nearby atolls in the Tuamotu Islands (Stoddart 1969b; Guilcher *et al.*, 1969; Pirazzoli and Montaggioni, 1985, 1986). The nature of this corrosion has been investigated by Hodgkin (1970) in the emerged coral reefs of the Langkava Islands and Samporia Bay in Malaysia, where deep notches and residual mushroom stacks are common. Hodgkin has shown that these erosional features are due to the action of endolithic algae, which are forced to penetrate deeper into the rock as they are rasped by chitons and other browsing molluscs, the boring sponge *Cliona*, the bivalve mollusc *Lithophaga*, and other boring organisms. In other words, the intertidal and supratidal erosion leading to such notches in mainly *bioerosion*, which is by no means restricted to coral areas since it has been also studied in detail on the calcareous shores of the Mediterranean Sea (Schneider, 1976; Le Campion-Alsumard, 1979; Dalongeville, 1986). However, the studies carried out by Trudgill (1976) at Aldabra Atoll in the Indian Ocean have indicated, in addition to bioerosion, solution by sea-water, a process that has not been generally acknowledged in recent times.

6. INFESTATIONS OF REEFS: *ACANTHASTER PLANCI*

This 'plague' of coral reefs is a starfish, also called the Crown of Thorns, which kills corals as it feeds on them (Vine *et al.*, 1973; Bradbury *et al.*, 1985). It is now widespread throughout the Indo-Pacific region, including the Red Sea, but has not been recorded from the Caribbean. It has been known since the eighteenth century, when it was named by Linnaeus, and its habits have been long documented. Although it is reported as common in a number of scattered areas, such as Christmas Island, the Solomon Islands, Java, and known to have various impacts on corals in many parts of the South West Pacific, its predations have drawn special attention to the Great Barrier Reef since the nineteen-sixties, when reefs near Cairns (lat. 17°S) became infested. Subsequently *Acanthaster planci* was reported from other areas between Townsville (19°30′S) and Cooktown (15°30′), and, in spite of apparent recoveries, renewed outbreaks have occurred in the central region of the Great Barrier Reef, and caused until 1984 heavy destruction of massive as well as branching corals (especially Endean and Cameron in Bradbury *et al*., 1985).

The recoveries have been explained by the fact that, even in areas which appear devastated on superficial examination, a significant and diverse stock of surviving corals remains. There is, however, little viable *Acropora* remaining, so that this genus is threatened (Done in Bradbury *et al*., 1985).

Several hypotheses have been put forward to explain the outbreaks of the Crown of Thorns on the Great Barrier Reef. According to Endean and Cameron, the most likely explanation is that humans have overcollected its predators, among which the giant triton shell *Charonia tritonis*, very much sought by tourists. The decrease of tritons would thus have allowed multiplication of *Acanthaster* in heavily visited areas. The same explanation has been suggested in Mauritius (Fagonee, 1985). If this is correct, it is one of the impacts of man on coral reefs, which will be examined more thoroughly in Chapter five. However, other authors, such as Frankel (1977), have expressed the opinion that the present infestations on the Great Barrier Reef are not abnormal, but are a phenomenon inherent to the ecology of the reef system.

Infestations by *Acanthaster* have also been described in Okinawa, Japan (Nishihira and Yamazato, 1974), where the density of their distribution has been found approximately proportional to human interference on reefs, an observation which would fit with Endean and Cameron's hypothesis on the Great Barrier Reef. But it is not certain that all these outbreaks had the same origin: in the Red Sea (Ormond and Campbell, 1974), where both *Acanthaster* and *Charonia tritonis* exist, there is no direct evidence that the latter is a predator of the former, while pufferfish and triggerfish have been seen eating actively the starfish.

7. IMPACT OF HURRICANES AND EARTHQUAKES

Hurricanes or typhoons (the second word being used in the China Seas) are extremely important because of the damage made in reefs by the winds and heavy seas which they generate (Stoddart, 1971a). They are practically unknown in the equatorial zone, and thus do not affect the reefs of Malaysia, Indonesia, Papua New Guinea, the Solomon Islands, or Kiribati (formerly

Gilbert), the Phoenix Islands, the Line Islands; but Tuvalu (formerly Ellice) and the Marshall Islands are subject to hurricanes; in the Indian Ocean, Kenya, the Maldives and Northwest Australia are exempt, as also the Red Sea and Northeast Brazil. They strike the tropical areas *sensu stricto* in both hemispheres, especially on the west side of the oceans. Their distribution month by month may be seen in *Atlas Okeanov, vol. 1*, p. 90–112 for the Pacific, and *vol. 2*, p. 91–113 for the Indian and Atlantic Oceans. In the Tuamotu and Society Islands, Southeast Pacific, they are quite rare, but can occasionally be exceedingly strong, such as the six which occurred in five months during the southern summer 1982–83, thought to be related to an exceptional oscillation of the Niño warm current, usually restricted to Ecuador and Peru coasts from December to April (Doumenge, 1983; Cane, 1986; Cane and Zebiak, 1985). Severe storms, not true hurricanes, damage reefs in Hawaii (Burch in Harmelin-Vivien and Stoddart, 1985).

Hurricanes include winds in excess of 120 kilometres per hour, with occasional gusts of over 300 kilometres per hour. Such winds produce extreme wave conditions, with a maximum height of 23.6 metres recorded during Hurricane Camille in the Caribbean in 1969, and associated storm surges to 5 metres above predicted tidal levels (Stoddart in Harmelin-Vivien and Stoddart, 1985).

Although the tsunamis, or waves generated by earthquakes and travelling over large distances across the oceans (especially the Pacific) are not often mentioned as a cause of destructions on reefs, they seem to be responsible for a part of the damage in the Tuamotu Islands (Bourrouilh-Le Jan and Talandier, 1985a).

The most extensive damage to corals is observed on the outer edges of reefs, which receive the direct impact of the waves. According to Laboute in Harmelin-Vivien and Stoddart (1985), the worst damage in Tikehau and Takapoto Atolls, Tuamotus, during the 1982–83 hurricanes, occurred under water on the slopes where, on the exposed sides, avalanches of blocks destroyed from 50 to 100 per cent of the living communities down to 40 metres at least. The main destructions were on the steepest slopes, the avalanches being not so dramatic where the slopes were more gentle. On Rangiroa Atoll in the Tuamotus, Bourrouilh-Le Jan and Talandier (1985a) have shown reef-edge fractures, with radii of curvature measuring 100 to 200 metres, from which blocks were detached. Blocks which were thrown up the reef flat (Fig. 31) were up to 20 metres long, their sizes decreasing from the outer edge towards the lagoon where blocks still measuring 2 metres were precipitated (see on Fig. 34 the general shape of the western Tuamotu atolls). Most of the blocks visible in 1985 on Rangiroa derived from the 1983 hurricanes, but this atoll has previously been struck by hurricanes in 1906, 1878 and earlier, and it is a place where tsunamis, recorded in old (sixteenth century?) native traditions, could have played a part. In the Caribbean, the observations made in Belize (formerly British Honduras) after Hurricane Hattie in 1961 (Stoddart, 1963, 1964a,b) indicate that 'immediately north of

Fig. 31 Mataiva Atoll, Tuamotu Islands. Giant detrital block pushed on the beach by a 1983 hurricane. The umbrella gives a scale. Still larger blocks, displaced during the 1983 events, can be seen in the Tuamotus (Photo A. Guilcher)

the storm track, up to 80 per cent of the reef corals disappeared on the barrier reef, presumably transported by wave action into deeper water'. In contrast with the Tuamotus, there was very little scattering of large blocks on the reef flats (Stoddart, 1963, p. 165). In all these Polynesian and Caribbean sites, beach rock has been remarkably successful in resisting erosion (same authors).

The recovery of coral growth on a slope after a hurricane varies according to sites. Two years after hurricane damage in the Tuamotus (Laboute in Harmelin-Vivien and Stoddart, 1985), *Pocillopora*, *Acropora* and *Porites* were among the species that had recolonized large areas down to 15 metres, and *Acropora* colonies of up to 60 centimetres in size were observed at a depth of 20 metres; but it will probably take several decades before the vigour of these biotopes recovers completely. At Funafuti Atoll in Tuvalu, observations made 8½ months after Hurricane Bebe (Baines *et al.*, 1974; Mergner, 1985) indicated that, beside regenerating segments of the few surviving massive coral colonies, young branched corals (*Acropora*, *Pocillopora*) were much more numerous, and were expected to have a decisive influence on the future zonation of the reef edge: this fits more or less what has been observed in the Tuamotus, but it is necessary to monitor changes that occur over a longer period. In Belize, recovery from 1961 to 1965 was slight: recolonization was limited to a few scattered *Acropora palmata*; *A. cervicornis*, formerly widespread, was still rare, and sponges, algae and gorgonians had made a more rapid recovery than corals (Stoddart, 1971a).

As a consequence of erosion, accretion of elements smaller than huge blocks occurs during hurricanes behind the reef edge. At Funafuti (Baines *et al.*, 1974), Hurricane Bebe built on the southeastern side of the atoll a rubble rampart 19 kilometres long and up to 4 metres high, with a calculated volume of 1.4×10^6 m^3 and a mass of 2.8×10^6 tons. Before the building of this storm beach, a 'hurricane bank' had been formed behind, in contact with the shore of the vegetated island. As its name implies, it is attributed to former hurricane(s), but it was displaced by Hurricane Bebe in some areas. In Rangiroa, the former beaches retreated lagoonwards, but not far since they were anchored on old beach rock. The 1958 hurricane at Jaluit Atoll, Marshall Islands (Blumenstock, 1961) built gravel tracts, shore ridges or ramparts, and gravel sheets or blanket deposits, but these were much smaller than the huge rampart built at Funafuti.

In Belize, a number of sand cays disappeared completely during Hurricane Hattie, 1961 (Stoddart, papers cited) (for sand cays in general, see Chapter four, section 3.6). These cays were generally vegetated, most were long-lying, all but one were predominantly sandy, and all were small. In many areas that coral sand cays are often swept by hurricanes, another example being off Tuo in the New Caledonia barrier reef (Guilcher, 1968 p. 190), where Ain Cay, which was wooded, was destroyed during a cyclone in 1932. In such cases, remnants of beach rock may persist to commemorate the outlines of the lost cay.

Even where the islands are not destroyed, their vegetation is damaged to various degrees. There are large differences between islands which have kept their natural cover, and islands with coconut plantations. In Belize, Stoddart (1964b, 1971a) found that on islands with natural littoral thicket, or with coconut keeping a dense undergrowth, Hurricane Hattie was unable to destroy more than an outer fringe of vegetation; but where coconut plantations had no undergrowth, the trees were easily removed by wind, waves, and erosion of soils. Subsequently the shrub vegetation close to the coast made a remarkable recovery, but tree growth was slower. Nevertheless in the 1962–65 period newly planted coconuts had grown to heights of 1–3 metres. More rapid revival was observed in the Tuamotus: at Mataiva Atoll for example, two years after the 1983 cyclones. It is certain that islands with vegetation left in natural state resist hurricane damage better than those with plantations; but as coconut trees are the only possible commercial cultivation on atolls, they are necessary on many of the islands.

From observations made so far after hurricanes, it seems that the recovery of the destroyed coral communities is diverse, and that, in many cases, one kind of system can be replaced by another. For example, as Stoddart (in Harmelin–Vivien and Stoddart, 1985) pointed out, 'in the absence of herbivores (which have disappeared in a hurricane) a coral-dominated system may be replaced by an algae-dominated system, for an indeterminate period'. A hurricane may thus be the origin of a discontinuity in reef genesis.

Damage on coral reefs by earthquake was described near Madang, north-east New Guinea (Stoddart, 1972). Resulting from an earthquake of force 7.1 on the Richter scale, the damage varied considerably with species and growth form. It also varied over short distances, from less than 10 per cent to more than 90 per cent over 2 to 3 kilometres, and was greatest on reef slopes, especially in almost enclosed basins. It was generally less than that associated to major hurricanes, but, considering the type of destruction, it may be that recovery is more delayed. 'Such events must have a considerable influence on the structure and composition of reef communities . . . Earthquakes may favour rapidly growing branching species, forming colonies which are repeatedly destroyed but which rapidly regenerate'. Countries such as North Melanesia (Papua New Guinea, the Solomon Islands, etc) which are among the most seismically active in the world, are not struck by hurricanes, so that these two different kinds of damage are, partly at least, distributed over two different kinds of areas.

It remains that other coral communities, such as those living in the Chagos, the Phoenix, or Brazil, escape the two types of events. So that, according to a remark made by Stoddart (in Harmelin-Vivien and Stoddart, 1985), a distinction should be made, more than usually, between reefs which are repeatedly devastated, and those which have escaped such disasters. Would reefs such as those existing in the Tuamotus, the Caribbean or New Guinea differ markedly in their physical characteristics, population structures, and biotic networks, from reefs outside the hurricane and earthquake zones? At first sight, the differences between two such classes are not evident, but a programme of investigations might reveal in the former more permanent geomorphological features than the short-term damage so far described on reef surfaces and bordering slopes.

CHAPTER THREE

The origin of coral reefs

During the exploration of the oceans it had become clear that reefs are distributed into three main types: fringing reefs growing close to a higher land, barrier reefs separated from the land by a lagoon, and atolls consisting of a coral rim around a lagoon without any land in it other than low islets. Other types are variations on these three main types, with the possible exception of ridge reefs. Fringing reefs seemed simple in origin, but barrier reefs and atolls require particular explanations for their genesis. The theories that have been proposed will be reviewed in this chapter, more or less in historical order, with more detail on the more recent ones, since they take account of modern research and techniques which were not available in the nineteenth and the first part of the twentieth century.

1. DARWIN'S THEORY, COMPLETED BY DANA AND DEVELOPED BY DAVIS, AS AN EXPLANATION OF FRINGING REEFS, BARRIERS, AND ATOLLS

Darwin's theory is one of the most famous geomorphological theories ever proposed, and it may be interesting to recall how it was developed.

As pointed out by Stoddart (1962b), the idea that fringing reefs became converted into barrier reefs, and then into atolls, by slow subsidence of an island foundation, did not arise from the observation of Pacific and Indian Ocean coral reefs during Darwin's voyage around the world on the *Beagle* (1832–1836). In his Autobiography, Darwin wrote: 'No other work of mine was begun in so deductive a spirit as this: for the whole theory was thought out on the west coast of South America before I had seen a true coral reef. I had therefore only to verify and extend my views by a careful examination of living reefs. But it should be observed that I had during the previous years been incessantly attending to the effects on the shores of South America of the intermittent elevation of the land, together with the denudation and the deposition of sediment. This necessarily led me to reflect on the effects of subsidence, and it was easy to replace in imagination the continued deposition of sediment by the upward growth of coral. To do this was to form my theory of the formation of barrier reefs, and atolls' (Barlow 1958, pp. 98–99).

The observations themselves were made initially in November 1835, when sailing across what was then called the Low or Dangerous Archipelago, now the Tuamotus, and soon after visiting the Society Islands, where he examined particularly Tahiti and Eimeo, now Moorea. From making these observations he wrote:

'If we imagine such an island (as Tahiti or Eimeo), encircled by a coral reef separated from the shore by channels and basins of still water, after long successive intervals to subside a few feet, in a manner similar, but with a movement opposite to the continent of South America; the coral would be continued upwards, rising from the foundation of the encircling reef. In time the central land would sink beneath the level of the sea and disappear, but the coral would have completed its circular wall. Should we not then have a Lagoon Island (i.e. an atoll)? Under this view, we must look at a Lagoon Island as a monument raised by myriads of tiny architects, to mark the spot where a former land lies buried in the depths of the ocean' (Barlow (Ed.), 1933, p. 400).

He next examined the Cocos or Keeling Atoll in the eastern Indian Ocean, which was considered by him as type of atoll, and prompted him to expound his theory in the first chapter of his book as follows (with two figures, redrawn here as they were).

'Let us in imaginative place within one of the subsiding areas, an island surrounded by a "fringing reef"—that kind, which alone offers no difficulty in the explanation of its origin. Let the unbroken lines and the oblique shading in the woodcut (in Fig. 32–1, but with large dots instead of oblique shading) represent a vertical section through such an island; and the horizontal shading (In Fig. 32–1, with small dots instead of horizontal shading) will represent the section of the reef. Now, as the island sinks down, either a few feet at a time or quite insensibly, we may safely infer from what we know of the conditions favourable to the growth of coral, that the living masses bathed by the surf on the margin of the reef, will soon regain the surface. The water, however, will encroach, little by little, on the shore, the island becoming lower and smaller, and the space between the edge of the reef and the beach proportionally broader. A section of the reef and island in this state, after a subsidence of several hundred feet, is given by the dotted lines: coral islets are supposed to have been formed on the new reef, and a ship is anchored in the lagoon-channel. This section is in every respect that of an encircling barrier-reef; it is,

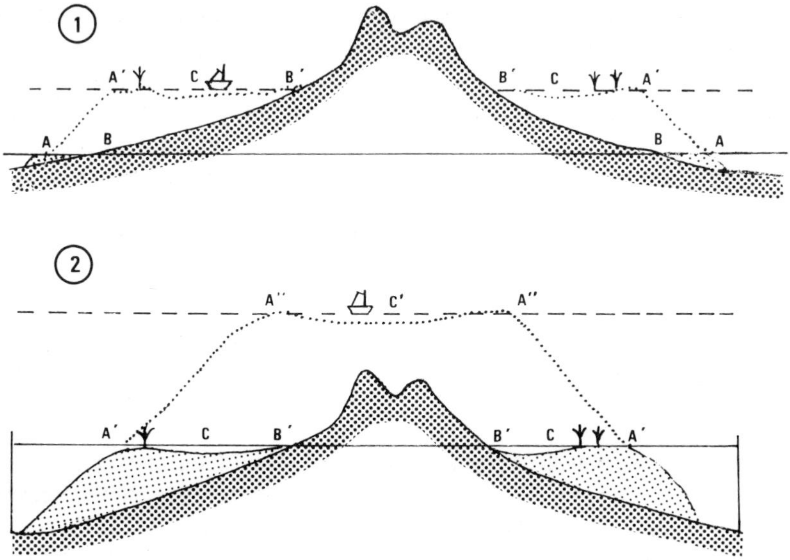

Fig. 32 Figures redrawn from Darwin (1842, Figs. 4 and 5), showing the formation of barrier reefs and atolls. Captions by Darwin. 1—AA: outer edge of the reef at the level of the sea; A′A′: outer edge of the reef, after its upward growth during a period of subsidence; CC: the lagoon-channel between the reef and the shores of the now encircled island; B′B′: the shores of the encircled island. NB: in this and the following woodcut, the subsidence of the land would only be represented by an apparent rise in the level of the sea.

2—A′A′: outer edge of the barrier reef at the level of the sea. The coconut trees represent coral islets formed on the reef; CC: the lagoon-channel; B′B′: the shores of the island, generally formed of low alluvial land and of coral detritus from the lagoon-channel; A″A″: the outer edges of the reef now forming an atoll; C′: the lagoon of the newly formed atoll. According to the scale, the depth of the lagoon and the lagoon-channel is exaggerated

in fact, a section taken E and W through the highest point of the encircled island of Bolabola; of which a plan is given in plate I, Fig. 5 (not reproduced here). The same section is more clearly shown in the following woodcut (Fig. 32–2) by the unbroken lines. The width of the reef, and its slope both on the outer and inner side, will have been determined by the growing powers of the coral, under the conditions (for instance the force of the breakers and of the currents) to which it has been exposed; and the lagoon-channel will be deeper or shallower, in proportion to the growth of the delicately branched corals within the reef, and to the accumulation of sediment, relatively, also, to the rate of subsidence and the length of the intervening stationary periods.

'As the space between the reef and the subsiding shore continued to increase in breadth and depth, and as the injurian effects of the sediment and fresh water borne down from the land were consequently lessened, the greater number of the channels, with which the reef in its fringing state must have been breached, especially those which fronted the smaller streams, will have become choked up by the growth of coral: on the windward side of the reef, where the coral grows most vigorously, the breaches will probably have first been closed. In barrier-reefs, therefore, the breaches kept open by draining the tidal waters of the lagoon-channel, will generally be placed on the leeward side, and they will still face the mouths of the larger streams, although removed beyond the influence of their sediment and fresh water; and this, it has been shown, is commonly the case.

'Referring to the following diagram (Fig. 32–2), in which the newly-formed barrier-reef is represented by unbroken lines, intead of by dots as in the former woodcut, let the work of subsidence go on, and the doubly-pointed hill will form two small islands (or more, according to the number of the hills) included within one annular reef. Let the island continue subsiding, and the coral-reef will continue growing up on its own foundation, whilst the water gains inch by inch on the land, until the last and highest pinnacle is covered, and there remains a perfect atoll' (Darwin, 1842, pp. 98–101).

Initially accepted with admiration, Darwin's theory, which received Dana's agreement (see later in this chapter), was subjected to many discussions and controversy at the end of the nineteenth century, and various theories were presented at that time, summarized by Davis (1928, p. 22) as follows (references to authors in Davis): '. . . whether the visible reefs have been formed by upgrowth as still-standing, aggraded submarine banks, as Rein first and Murray later suggested and as Wood-Jones still more recently argued; by outgrowth from still-standing foundations, the lagoon being excavated by solution back of the living reef, as Murray also proposed; by outgrowth on rising foundations, as Semper and Guppy believed, Semper having anticipated Murray in explaining lagoons by solution; by upgrowth on still-standing foundations completely truncated by abrasion, as Wharton proposed; or incompletely truncated, as Tyerman and Bennet long ago and as Guppy later suggested; or on platforms produced around still-standing foundations by subaerial erosion and submarine denudation, as Agassiz believed; or on still-standing foundations abraded by the lowered Glacial ocean and submerged by the rise of the Postglacial ocean, as Daly argues—for all these theories succeed in explaining the visible features of reefs that they were invented to explain. It is with regard to the invisible conditions and processes of the past that the theories differ'.

It must be pointed out that Darwin knew well, and said in his book, that many reefs around the world needed other explanations, generally complementary to his own general theory. He noted that in several areas there are elevated coral reefs. For example, at Savage Island, southeast of the Friendly Group (now called the Tonga Islands) (1842, p. 133), 'an ancient lagoon is now represented by a central plain: here we cannot doubt that the elevatory forces have recently acted. The same conclusion may be extended, though with somewhat less certainty, to the islands of the Friendly Group . . . The surface of Tongatabou is low and level, but with some parts a hundred feet high; the whole consists of coral rock . . . On the New Hebrides (now Vanuatu), Mr. G. Bennett informs me he found much coral at a great altitude, which he considered of recent origin . . . In the East Indian archipelago, many authors have recorded proofs of recent elevation' (Darwin, 1842, p. 133–134). He accepted evidence on other raised reefs, as in Mauritius, Timor, New Guinea, and the Sandwich Islands (now Hawaii).

Concerning Tongatabou Island, Darwin's view has been supported by more recent and more accurate work. Taylor and Bloom (1977) described terraces at 40–80 metres and 110–140 metres, indicating uplift and tilting on this island. The uplift has occurred at an average rate of at least 3.3 metres/100 ky in post-Miocene times, but has now ceased, or become very slow.

In many respects, it would be wise for modern scientists to read Darwin's book carefully, for they would find many facts and opinions which are often presented as recent discoveries: for example, the fact that passages through barrier reefs and atoll rims occur on the leeward side, but are closed on the windward side, where they are thought to have formerly existed, is mentioned by Darwin (1842, p. 100 and 147). In the Red Sea also, it will be shown later in this book (Chapter four, section 5) that essential features of coral reefs were mentioned by Darwin in 1842 on the basis of informations he had collected from sailors.

It is curious, as Davis (1928, p. 44) has pointed out, that Darwin failed to cite the embayed shorelines of islands surrounded by barrier reefs in support of his theory of subsidence. It was Dana who drew attention to that evidence, and so he must be mentioned with Darwin as a pioneer of the explanation of barrier reefs and atolls. Dana explored the Pacific reefs in 1839, just four years after Darwin, as a member of the United States expedition under Captain Wilkes. The radial valleys of Tahiti suggested to him that 'sunk to any level above that of five hundred feet, the erosion-made valleys of Tahiti would become deep bays, and above that one of one thousand feet, fiord-like bays, with the ridges spreading in the water like spider's legs' (quoted by Davis, 1928, p. 46). It is also surprising that, in spite of the fact that Dana's ideas in this respect were documented in the three successive editions (1872, 1874, 1890) of his classic book *Coral and Coral Islands*, the value of embayments as evidence of subsidence was not appreciated by other scientists interested in coral reefs, such as Guppy (1888) and Agassiz (1903b).

'Darwin—Dana—Davis, the three Ds. The subsidence theory of coral barrier reefs and atolls will always be associated with these three men. Darwin, the originator, Dana, the strengthener, and Davis, the skilful champion. It is a curious fact that all three were born on February 12th, Darwin in 1809, Dana four years later, and Davis in 1850' (Hoffmeister, 1940).

The antecedent-platform theory, which was proposed by Hoffmeister and Ladd (1944), may be mentioned here as an extension of Darwin's theory with inclusion of further data. It had evolved progressively through the work of authors like Wharton, Murray, Vaughan, Gardiner, Semper, Rein, Wood

Jones, details of their discussions being given by Davis (1928). According to these ideas, corals would have grown on any basement providing favourable conditions: old volcanic islands truncated by abrasion; submarine volcanoes not truncated; deep submarine foundations of any origin, raised by sedimentation until they would become shallow enough to allow coral growth; or uplifted or downfaulted rocky basements affording the same possibilities. Later on, reef building would have continued during slow subsidence, as in Darwin's theory, particularly in Indonesia and in the Central Pacific. The subsidence may be tectonic, or isostatic, or due to compaction in the deep-sea floor. Such an evolution would have not occurred in the Western Pacific area, where reefs have been commonly uplifted. These differences in evolution between parts of the Pacific were suggested at a time when the notion of sea-floor spreading had not yet been developed, but the idea proved, later on, to be fruitful and even premonitory.

2. SEA-LEVEL SHIFTS AND THEIR EFFECTS ON REEFS

When the effects of Pleistocene glaciations and deglaciations on sea-level were discovered (and some authors mentioned them long before the end of the nineteenth century), it appeared necessary to take account of them in explanations of the coral reef types, especially as the more elaborate types apparently required a long time to be built. But knowledge of the Quaternary shifts in sea-level has been considerably improved in the course of time, as the knowledge of the growth of corals during the last, Holocene transgression.

Daly is the outstanding name in the early phase of these studies. His contributions to the study of coral reefs in relation to sea-level changes extended over a long period (Daly, 1910, 1934, 1948). His basic idea concerned the formation of the large barrier reefs and atoll lagoons, and may be summarized as follows. The depth of these lagoons is thought to be generally nearly uniform and flat at approximately 70–80 metres. It is suggested that they were abrasion platforms, now covered by a veneer of debris and calcareous algae; they would have been, perhaps not completely shaped, but at least levelled, during a low, interglacial sea-level. The fall of sea temperature at that time would have killed the majority of the corals, and even if some of them survived (Daly in his 1948 paper accepted evidence, from Dutch work on Indonesian coral reefs, that they never disappeared in the warmest seas), their growth was hindered by increased turbidity. As temperature rose in Late-Glacial and Post-Glacial times, corals would have spread from their Late Pleistocene sanctuaries, and recolonized the platforms, their ensuing upward growth accompanying the rising sea-level.

Nugent (1946) published a description of the coral reefs in the Gilbert, Marshall and Caroline Islands, which was in support of Daly's theory.

Beside the basic fact that sea-level variations have occurred, which is undoubted, Daly's theory has been heavily criticized. The assumption of widespread influence of turbidity during low sea-level phases is very contestable. As has been said (Chapter one), turbid waters are unfavourable to coral growth, but in several instances corals (or species of corals) are able to grow in such conditions. Shepard (1963, p. 353) quoted Ladd and Hoffmeister 'who have called attention to the existence of reefs off Rewa River in the Fiji Islands in spite of the tremendous amounts of silt brought to the sea . . . Kuenen also refers in Indonesia to floods covering a reef with silt without killing it'. He mentions similar cases in Hawaii and elsewhere.

Even more important is the fact that the depth of lagoons is not as uniform as Daly assumed. Its diversity was already recognized when Davis wrote his book (1928, p. 92–97) and has been checked by more recent work. Most soundings show that lagoon depths are often very irregular. The fine charts of four of the Marshall Islands, published by Emery *et al.* (1954), are quite clear in this respect. From Japanese surveys, Tayama (1952, p. 248) has concluded that average depths of the lagoons are 27 metres in the Palau Group, 41 metres in the West Caroline Group, and 63 metres in the East Caroline Group. In addition, there are some extremely shallow lagoons, as at Christmas Atoll, Line Islands, and Elizabeth Atoll, between Australia and New Caledonia, each of which having a lagoon only a few metres deep (Slater and Phipps, 1977; Valencia, 1977). By contrast, a number of lagoons with depths of 150 and even 200 metres have been mentioned in the Moluccas (Kuenen, 1950, p. 455). Even if the depth of most atoll lagoons would be around 70–80 metres, this figure would not be indicative of low sea-level bevelling, since most recent estimates of sea-level lowering during the coldest period of the Wichselian (Wisconsinian, Devensian) have been around 125 to 130 metres. There have also been oscillations of sea-level during the Weichselian, with a temporary rise to slightly below present datum around 35 000 BP (Thom, 1973), with a deep fall afterwards. Similar shifts must have occurred during earlier glaciations. It thus seems unlikely that there have been sufficiently long stillstands at low levels to permit the bevelling envisaged by Daly.

A more valuable result from recent research is that the lowerings of the sea-level during Pleistocene glaciations have often determined a karstification of preexisting coral structures and a subaerial dissection of these structures by rivers, hence a diversification of their topography.

At Mayotte Island in the western Indian Ocean, the floor of the western part of the lagoon between the barrier reef and the central volcano has several enclosed depressions with steep sides, 60–70 metres deep, lying at approximately 20 metres below the surrounding lagoon floor (Fig. 85). The largest one measures 1800 by 2800 metres, and includes pinnacles 15 to 20 metres high. These features have been interpreted as the results of a subaerial karstification, the pinnacles being needles of a small Kegel Karst, possibly coated by thin coral growth added since the postglacial transgression (Guilcher, 1965b; Guilcher *et al*., 1965a).

At Clipperton Atoll, off Mexico in the Eastern Pacific, the southeastern part of the lagoon includes a narrow pit with vertical sides, more than 36 metres deep, surrounded by shallow flats (Sachet, 1962, p. 51–52). Diving in

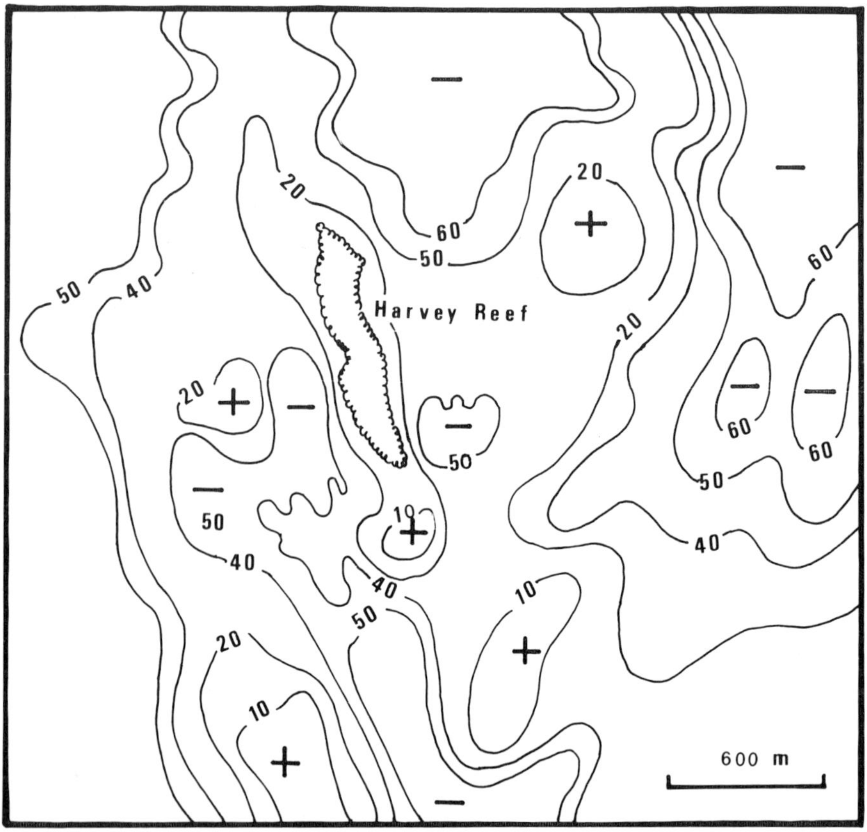

Fig. 33 Generalized contours (in metres) around Harvey Reef, off Port Sudan, Red Sea, showing several enclosed depressions, probably of karstic origin. After Braithwaite (1982). Harvey Reef itself, however, may be a ridge reef (cf. Figs. 137 to 142)

this pit by Cousteau's team, and further measurements by Ehrhardt (1976), revealed that its water has a high content of hydrogen sulphide, an evidence for its confinement. It is undoubtedly a result of subaerial karstification.

Off the Sudanese coast in the Red Sea, the Harvey Reef, 14 kilometres SSE of Port-Sudan (Fig. 33), which runs north and south and rises to the surface, is surrounded by enclosed depressions, 50 to 60 metres deep, their margins being only 15 to 20 metres in depth. The general pattern is extremely irregular and consists of a system of pinnacles and sharp ridges (Braithwaite, 1982). 150 kilometres north of Port-Sudan (same author), the Baraja Reef also includes enclosed depressions and irregular pinnacles which rise to within 15–20 metres of the water surface across the bank. These landforms point to a pre-Holocene dissection of previous reefs, the pits approaching perhaps 60 per cent of the area covered. The floors of the pits are now blanketed with recent sediments, while the walls of the intervening ridges have become sites for reef growth. Drowned features that are presumably karstic were also mentioned from near Port-Sudan by Schroeder and Nasr (1983). They noted that the basins or pits provided environments of deposition for very fine grained material, including a relatively high proportion of non-carbonate particles, and that such very low energy environments were located in close vicinity to reefs exposed to strong surf.

In the Bahamas, which may be classified as a special type of coral reef (see Chapter four, section 5), drowned karst has been described by several authors, especially on and around Andros Island, and other karstic features occur on reefs off Belize in the Caribbean (Benjamin, 1970; Bourrouilh, 1974; Stoddart, 1962a, p. 53; Purdy, 1974, p. 14; Cousteau and Diole, 1973; Dill, 1977). At Andros, numerous pits, known as 'blue holes', are circular submerged dolines, connected beneath the reef with the eastern submarine scarp of the Great Bahama Bank. Some are small, others reach 60 to 80 metres in diameter, sometimes with overhanging sides. The Holocene transgression resulted in muddy sedimentation with algal components in the shallow dolines. Off Belize, fossil stalactites were observed in caves down to 122 metres. These caves showed horizontal enlargements of joint patterns, which form tunnels at depths of 12, 18, 28, 40, 45, 55 metres, considered to indicate stillstands of sea-level during the Pleistocene regression. Sand waves are evidence that strong currents are now moving sediments down to 55 metres in open spaces.

The calcareous rocks of Late Cretaceous to Eocene age which outcrop on the continental shelf off Northern Florida are not former coral reefs, but they show submarine dissolution features in the form of sinkholes (Popenoe *et al.*, 1984) which are interesting in comparison with those found in reefs, since they help to understand the subaerial weathering on various limestones during phases of low sea-level. These sinkholes are incised abruptly into the sea floor, particularly the Red Snapper Sink, which occurs 40 kilometres offshore at a depth of 27 metres, and has been investigated down to 100 metres. The major period of solution occurred in the Late Oligocene and Early Miocene, but it continued during Pleistocene low sea-level phases. As in the Red Sea off Sudan, these holes are being infilled with shelf sands, indicating that they are no longer being shaped by solution processes.

The effects of processes at work during low sea-level phases on the general shape of atolls, particularly in the Pacific, will be discussed later in this chapter (section 5).

Beside the subaerial karstic effects of water circulation resulting in the sinkholes presently submerged, the phases of low sea-level led to the cutting of valleys now drowned beneath lagoons and through atoll rims and barriers. Four such valley patterns were reported in the Mayotte lagoon (Fig. 85), connected with passages across the barrier (Guilcher, 1965b; Guilcher *et al.*, 1965a). The most remarkable one, in the northeastern part of the lagoon, is 60 to 80 metres deep and ends in the Mzambourou Passage. Another one, 30 to

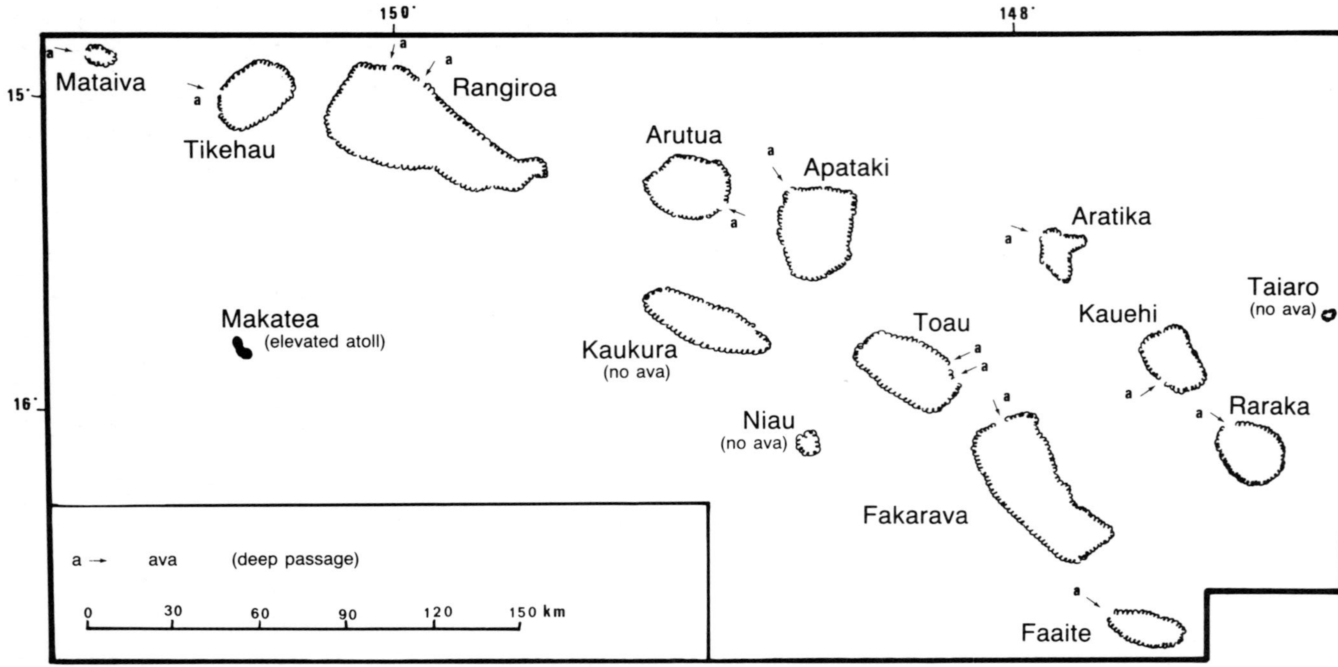

Fig. 34 Northwestern part of the Tuamotu Islands, showing the location of the avas (deep passages, available for ships). The avas are generally located on the leeward side of the atolls, except when one atoll is sheltered from the trade winds by another

40 metres deep, is connected with the Longogori Passage in the east: this sinuous passage (Fig. 50), 4 kilometres long, only 300 to 400 metres wide, is 50 to 60 metres deep, and has only occasional, discontinuous sandy deposits on its bottom, and steep slopes on either side. The unusual lack of sedimentation in such Pleistocene erosional landforms in the Mayotte lagoon seems due to strong scouring currents produced by the large tidal range, which reaches 4 metres at spring tides, a rather high figure for tropical countries.

Similar features are found at Guadeloupe Island, in the Caribbean (Guilcher and Marec, 1978) (Fig. 94), where passages across the barrier reef of Grand Cul-de-Sac Marin are connected with former valleys in the lagoon behind; although the depths are not so large as at Mayotte (35 metres across the barrier), the origin is the same.

The Tuamotu atolls in the Pacific are connected with the open sea by two types of passes: the shallow ones known as *hoa*(s) which will be dealt with in Chapter four, section 4.6, and the *ava*(s) which are much less numerous, deeper and navigable by ships (Fig. 34). The latter (Guilcher, 1966, 1979b) are certainly remnants of fluviatile Pleistocene valleys that drained from the emerged lagoon into the lowered sea. They are generally located on the leeward side of atolls (or the barrier as in the case of Bora Bora in the Society Islands), in sectors not exposed to the trade wind swell, except when one atoll is sheltered by another, in which case they are found where the surf action is minimal. The location of the passages thus seems to be determined by swell, a feature independent of sea-level lowering. The same leeward situation of passages may be seen in the Marshall and Caroline Islands, where they lie mostly in the southwest at the opposite side of the trade wind action in the northern hemisphere (Wiens, 1962, p. 33–35). Guilcher (1979b) thought that other valleys existed all around the atoll rims and barriers during low sea-level phases, but that the surviving ones are those which were sheltered from sedimentation which has closed the passages on the windward side.

Such an explanation, previously suggested by Darwin (1842, p. 100: 'on the windward side of the reef . . . the breaches will probably have first been closed'), receives confirmation from recent high resolution seismic investigations carried out on the rim of Mururoa Atoll in the Tuamotu Islands (Ruzie and Gachon, 1985, p. 388), which have discovered a series of small gorges cut into reef formations and subsequently filled by sand.

Similar landforms are reported from frameworks other than atolls, such as those situated in the northern region of the Great Barrier Reef, where Orme (1978) and Orme *et al*. (1977, 1978a,b) have used high resolution profiles to demonstrate the existence of submerged valleys, some infilled, while others remain open, without recent sedimentation. Fig. 35 gives examples of both types, the open one in the inner approaches of Cook Passage between two outer reefs, where currents have kept the bottom free of sediment, while the infilled one is more sheltered, and has retained deposited material. Orme (1978, p. 35) noted that these submerged channels can be of various ages, but the fact stressed here is that they always bear testimony of former low sea-level phases.

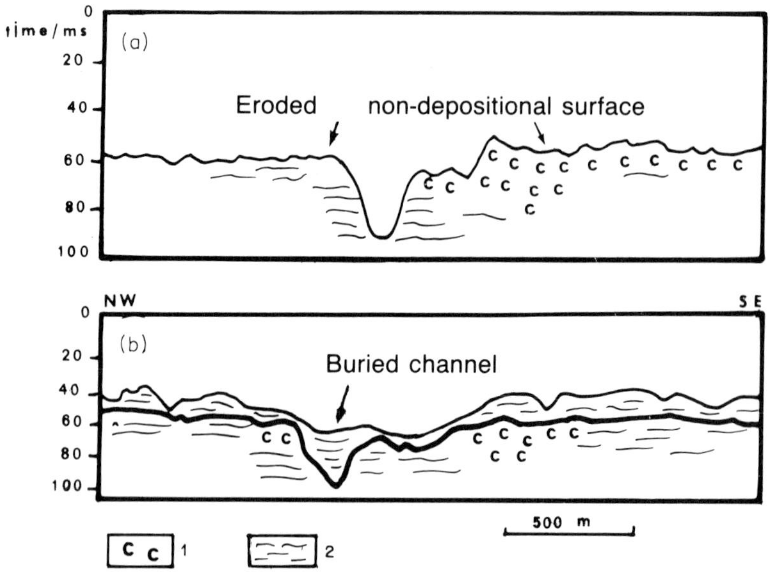

Fig. 35 Two NW–SE high-resolution seismic profiles across the Great Barrier Reef lagoon off Lizard Island. (a) Just behind Cooks Passage, showing the Pre-Holocene channel still open. (b) Farther into the lagoon, showing the same channel buried by Holocene sedimentation. 1: reef-rock; 2: bedded back-reef deposits; heavy line on B: Pre-Holocene surface. After Orme, Flood and Sargent (1978)

Sometimes the submerged channels have been overfilled by sediments, so that they stand as positive relief above the general bottom topography. This overfilling has happened in the central sector of the Great Barrier Reef (Hopley and Murtha, 1975; Johnson *et al.*, 1982) (Fig. 36), where these features have been attributed to thick backfilling by deltas during abandon-

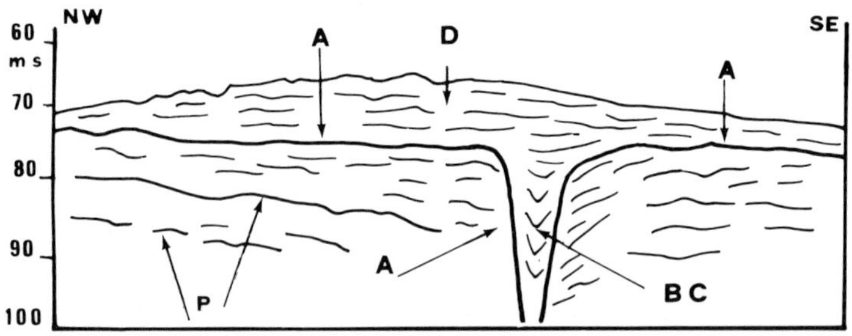

Fig. 36 Continuous seismic profile across the Great Barrier Reef lagoon off Townsville, showing a Holocene delta (D) built over a Pleistocene buried channel (BC). A: Pre-Holocene erosional surface; P: Older Pleistocene erosional surfaces. After Johnson, Searle and Hopley (1982)

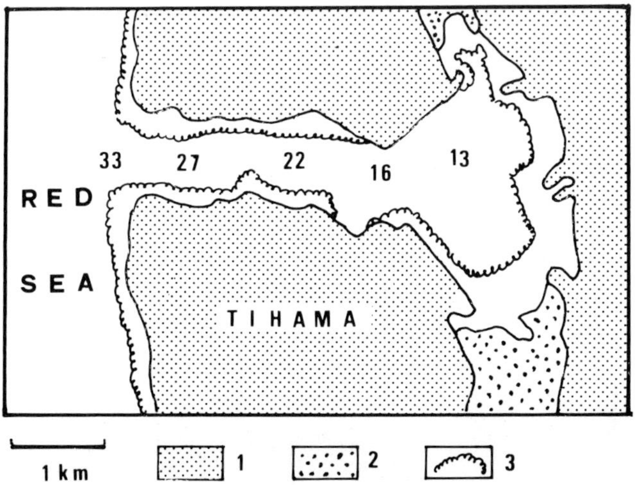

Fig. 37 Sharm Jubbah, on the Red Sea coast of Saudi Arabia at 27°30′N, drowned valley with Holocene coral growth on its sides. 1: Tihama (coastal plain); 2: alluvial deposits in wadis; 3: fringing coral reefs; Depths in metres

ment causing deposition in the lower reaches of the river. 'Hence modern shelf morphology may not be an accurate guide to the positions of Pleistocene channels across the shelf'.

Another result of sea-level changes is the occurrence of *sharms* or *marsas*, quite frequent on the Red Sea coasts of Saudi Arabia, Sudan, and the Sinai Peninsula in Egypt (Crossland, 1907–1915b; Schmidt, 1923; Rathjens and Von Wissmann, 1933; Guilcher, 1955, 1958 p. 156, 1979, 1982a, 1985b; Sestini, 1965; Mergner, 1967; Gvirtzman *et al.*, 1977; Dalongeville and Sanlaville, 1981). A sharm (Arabic plural: shurum) (Figs. 37 and 38) is a long, narrow, deep inlet, sometimes meandering, with corals growing on its shores, and sediments washed in from the wadi(s) at the inner end: Sharm Abhor, close to Jeddah, is a fine example (Guilcher, 1955, p. 64–65). Sharms are valleys that were incised by rivers during Pleistocene low sea-level phases, and may be considered as a special kind of ria formed in an arid coral environment. Various stages of infilling are found, and sometimes the only indication of the former sharm is a re-entrant in the fringing reef on the coast, as on parts of the Gulf of Aqaba or Elat, on the east coast of the Sinai Peninsula. In Sudan, where they have been cut into old, Pleistocene fringing reefs, sharms often display inner widenings behind a narrow entrance, because the valley cut during the low sea-level phase was more easily carved in the loose sediments of the former boat channel than in the more resistant outer edge of the emerged Pleistocene reef, structured by coral colonies in position of growth. The harbours of Port-Sudan and Suakin utilize such

Fig. 38 Southeastern coast of Sinai Peninsula, Red Sea. Remnant of sharm in fringing reef (Photo A. Guilcher)

sharms. They are not confined to the Red Sea. On the coast of Kenya (Bird and Guilcher, 1982), the narrow valleys cut across the emerged Pleistocene reefs (Battistini, 1969) have formed inlets that widen landward, as at Mombasa Harbour, Shimo and Kilifi. These are equivalents of the Sudanese marsas, with a new fringing reef and boat channel added as as a sequel to the Holocene marine transgression. In the Pacific, Erromango Island in Vanuatu (New Hebrides) has several sharms, particularly at Ipota drowned valley in Cook Bay on the east coast, which has been cut into an emerged coral reef lying at 3 to 4 metres above present sea-level. Fig. 39 shows a rather similar but smaller feature at Haiti, Caribbean.

Fig. 39 Haiti, Hispaniola Island, Caribbean. North coast near Cap Haitien at Ducroix beach. A channel cut by a river into the fringing reef during a Pleistocene low sea-level has been submerged by the Holocene transgression (Photo A. Guilcher)

The oscillations of high and low Quaternary sea-levels, which had such important impacts on reef building and erosion, have been much studied in recent decades. Unfortunately, the facts appear more complicated in 1987 than it was thought around 1940–1950.

Concerning former high sea-levels, the penultimate (Holsteinian–Hoxnian–Yarmouth: the terminology differing according to countries), and the last one (Eemian–Tyrrhenian–Ipswichian–Sangamon) are the best known, but several earlier ones have occurred (Fairbridge, 1961; Guilcher, 1969). There were secondary, minor oscillations of sea-level during the Eemian, which has lasted at least 50 000 years, and the maximum level attained by the sea in the Eemian is subject of much discussion. According to some authors, it would not have risen more than a few metres above present datum, but others think that it could have reached 15 or 20 metres without any significant tectonic influence. Evidences of higher sea-levels, before or during the Holsteinian, could be ascribed to the uplift of coastal land margins rather than to movements of the sea itself. From evidence in the southwestern Indian Ocean, Battistini (1972) has suggested that the Pleistocene sea never rose more than a few metres above the present level. The occurrence of very high emerged coral reefs in regions such as Papua New Guinea and Vanuatu are obviously related to large tectonic uplifts and will be discussed in section 7 of this Chapter. But even on coasts where there has been little, if any, uplift, emerged Pleistocene coral reefs are frequently exposed and have a geomorphological significance, as those which constrict former boat channels in Sudan and in Kenya.

In New Caledonia, where a drilling (Fig. 40) has been made down to the pre-coralline basement at 226 metres below the present sea-level in the barrier reef at Grand Ténia (Coudray, 1976, p. 119–175, and 1977), stratigraphic evidence of three successive reefs backed by sandy lagoonal deposits has been identified below the present reef. They have been tentatively related to Sangamon, Yarmouth, and Aftonian. They lie well below present sea-level because of the subsidence that accompanied the building of the barrier reef. The contact between the Pleistocene and Holocene reefs is at a depth of 11 metres (see also Thomassin and Coudray, 1981). Drilling in the Great Barrier Reef of Australia, down to 182 metres at Michaelmas Cay in 1926, and to 222 metres at Heron Island in 1937 (Fairbridge, 1950, p. 382-390), did not provide much information on the succession of Pleistocene reefs, but the unconformity at the base of the Holocene in that area is at depths between 5 and 20 metres according to various drillings (Hopley, 1982, p. 198). For the Great Barrier Reef area before Pleistocene, see this chapter, section 3. Drillings in atolls are examined in this chapter, section 4.

For the period of sea-level rise during the end of the Weichselian–Wisconsinian and the Holocene, a very considerable amount of new data has been provided by Project 61, *Sea-Level Movements During the Last 15 000 Years*, under A. L. Bloom's direction in the UNESCO International Geological Correlation Programme. The results, included in Bloom's *Atlas of Sea-Level Curves* (1977–1979), have been widely discussed (see Kidson, 1982; Hopley, 1982 p. 169 and 387; and Hopley and Pirazzoli, 1985).

Sea-level history in this period is summarized in curves, which can be classified into two main types: those showing a continuous rise in sea-level, with a rate diminishing in time but without oscillations; and those with oscillations, showing a general rise with alternating transgressive and regressive phases. In the positive phases of the oscillating type, sea-level surpasses frequently its present position: six times from 5700 to 1000 BP in the curve presented by Fairbridge (1961). Similar results have been put forward for coral areas, with Holocene sea-levels rising to more than 3 metres (5 metres in Brazil) above the present level. Coral areas are very favourable places for such investigations, because dead corals, provided they are in position of growth (that is, not coral heads thrown by storm waves to another level) give precise evidence of former sea level at low ordinary tides. Such emerged Holocene reefs, ranging in age from 1200 to 4000 or even 6000 years, have been reported from the Society, Tuamotu and Cook Islands, Midway and Kure Islands in Hawaii, Ifaluk in the Carolines, Jarvis and Starbuck in the Line Islands, Enderbury in the Phoenix, New Caledonia, etc (Stoddart, 1969b; Guilcher *et al*, 1969; Pirazzoli and Montaggioni, 1985, 1986; Scoffin *et al*., 1985; Ladd *et al*., 1970; Tracey and Ladd, 1974; Coudray, 1976, etc),

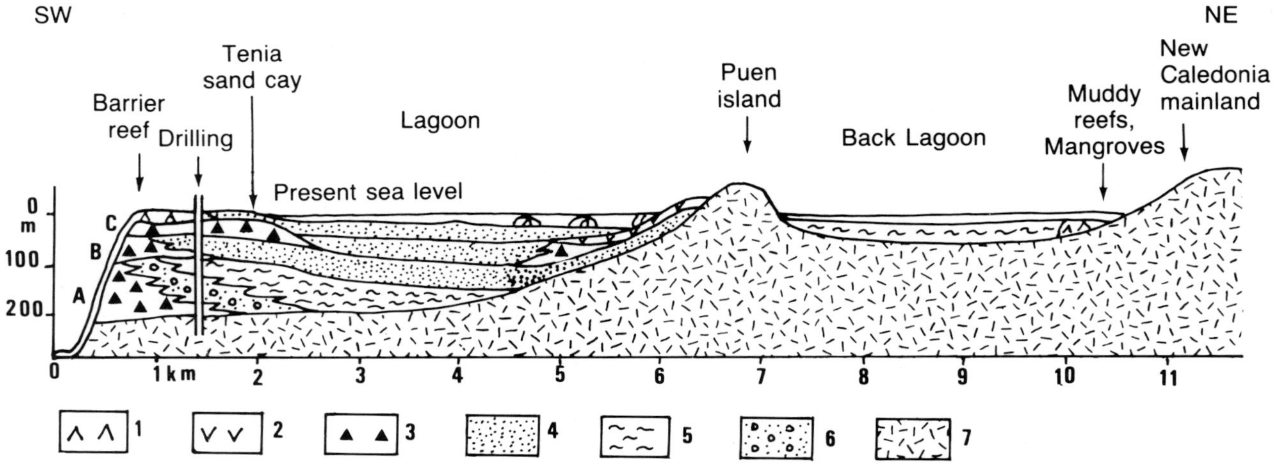

Fig. 40 Drilling through the New Caledonia barrier reef off Saint Vincent Bay near Noumea, and the interpretative cross section, after Coudray (1976) (Singer Polignac Expedition). 1: Holocene reefs; 2: Late Pleisocene aeolianites; 3: Pleistocene reefs (A, B, C); 4: Lagoon sands; 5: Lagoon muds; 6: Pleistocene back reef deposits; 7: Pre-Oligocene basement

have been considered as evidence of Holocene sea-levels higher than the present. Corrosion notches have also been invoked as indication of higher Holocene sea-levels in parts of these archipelagoes and elsewhere. Montaggioni and Pirazzoli (1984) even suggested that reef conglomerates, including coral heads thrown by waves, allow deductions about the former heights of the zones in which they were formed. Although recent tectonic disturbances have occurred (e.g. at Suwarrow Atoll, Cook Islands, according to Scoffin *et al.*, 1985), Holocene emergence(s) of reefs by a lowering of sea level has certainly occurred.

But an essential fact resulting from many recent works is that these conclusions cannot be extrapolated outside the areas in which they are established; in other words, that (Kidson, 1982) 'the search for a universal eustatic curve must be regarded as over; regional differences in response to change in the geoid mean that eustatic sea level curves can have only regional validity', and, on the other hand, that 'no part of the earth's crust can be regarded as wholly stable'.

Glacial isostasy, consisting of the plastic depression of the earth's crust under the loading of the ice sheets, with a slight complementary elevation near the edge of the ice, and rebounds after the melting of the glaciers, does not concern coral reef areas. More recently, however, there has been a growing appreciation of hydro-isostasy, which is a crustal response, particularly in shelf areas, to the differences in loading by the sea water alternately kept and released by the glaciers. Hydro-isostasy applies not only in glaciated areas, but also in coral areas, and its effects are expected to vary around the globe. The gradient of the sea-level curve during the Holocene may thus be an indicator of mantle viscosity (Bloom, 1967; Chappell, 1974b, 1983; Lambeck and Nakada, 1985), and such considerations could provide a solution to the discussions on Holocene sea-level changes in the Great Barrier Reef area.

More widely, a 'geoidal eustasy' has been proposed by Mörner (e.g. 1976), resulting from changes in the geoid, and thus different from the eustasy of glacial origin. Satellite surveys have confirmed that bumps and depressions exist in the geoid surface, probably related to density variations within the earth, and it is likely that these highs and lows migrate and vary in dimensions. A multidisciplinary approach in sea level research has been initiated by IGCP Project 200, *Sea Level Correlations and Applications*, under the chairmanship of P. A. Pirazzoli, 1983–1987. These trends in research throw a new light in the interpretation ofHolocene sea-level curves, and hence of the growth of the coral reefs since the end of the last glaciation. Clark *et al.* (1978) used a numerical model to define six contrasting sea-level zones in the world, of which four are in areas of coral reef growth. An example may be quoted from the Brazilian coast, which has coral reefs, where Holocene sea-levels 2.3 to 5.0 metres higher than the present have been recorded, and tentatively related to geoid surface deformations (Martin *et al.*, 1985). Fig. 42 shows a set of contrasting Holocene sea-level curves for four coral reef areas.

3. REEF GROWTH IN RESPONSE TO THE HOLOCENE TRANSGRESSION

The growth of reefs during the Holocene marine transgression has been studied in several areas during recent years, in relation to local sea-level curves, and deserves special attention as a consequence of shifts in sea-level.

Numerous American studies on reefs in the Caribbean and the nearby tropical Atlantic, using diver-operated drilling to investigate the internal facies, have provided information on the various ways in which these reefs responded to the sea-level rise. The basement consists of Pleistocene fossil reefs, sometimes accompanied by lithified aeolianites, as on the southwest coast of Jamaica (Henry and Head, 1985). After detailed studies such as Adey's one on St Croix reefs (1975), and preliminary syntheses as by Adey *et al*. (1977), Neumann and MacIntyre (1985) have defined the three ways in which the Caribbean reefs have grown.

A first type (Fig. 43) consists of *keep-up reefs*, i.e. reefs which were able to

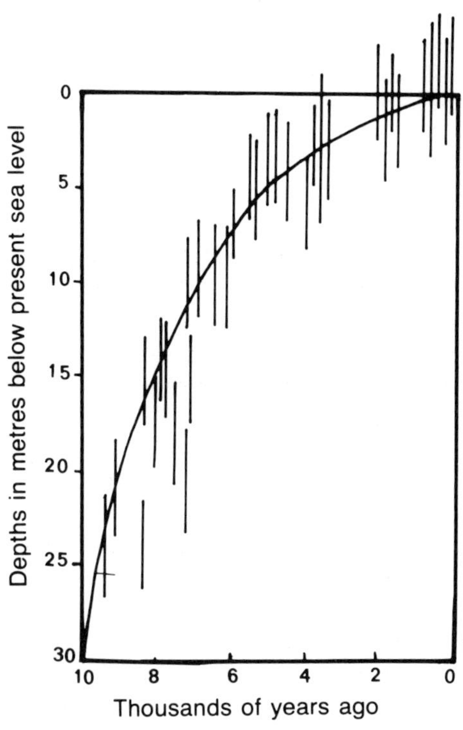

Fig. 41 Curve of sea-level rise based on *Acropora palmata*. Samples from Bahama, Florida, Martinique, Panama, Puerto Rico, St Croix. After Lighty, MacIntyre and Stuckenrath (1982)

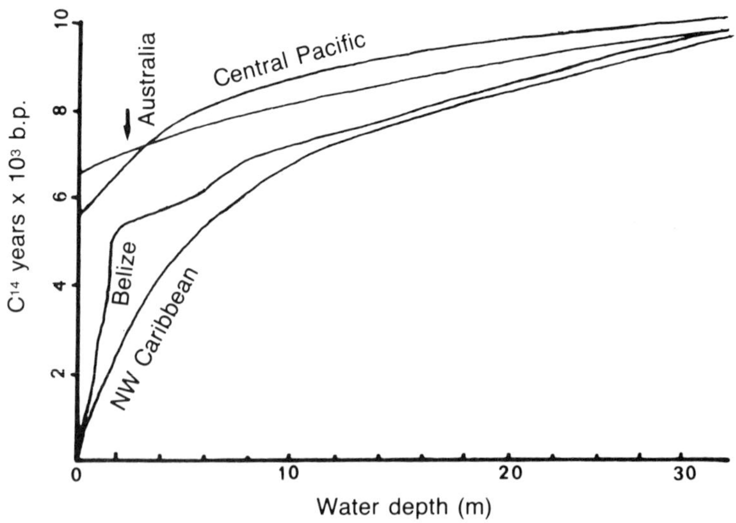

Fig. 42 Holocene sea-level curves for four selected coral reef areas. After Davis and Montaggioni (convenors) (1985)

maintain shallow, frame-building communities throughout the sea-level rise. 'Growth rate tracks rise rate'. These reefs, which occur on windward, narrow shelves, are voluminous, wedge-shaped, with layered isochrons. At the rate of sea-level decreased, they have spread laterally. Galeta Reef, off the Caribbean coast of Panama, is a good example. It began to grow at a depth of about 15 metres, and maintained an *Acropora palmata* facies until 2000 BP. The reef off Nonsuch Bay in Antigua is also a keep-up reef, with *A. palmata* prograding seaward during the last 3000 years. *A. palmata* framework is a very reliable reference for reconstructing the history of Late Quaternary sea-levels, owing to its restricted depth range (5 metres), the ease of obtaining uncontaminated samples, and its minimal compaction (Lighty *et al*., 1982) (Figs. 41 and 42).

A second type (Fig. 43) includes the *catch-up reefs*, which begin as shallow reefs and become deeper as rise rate exceeds accretion rate; then they grow upward and catch up with sea-level. They are found on protected shelves or back reefs. Their internal structure shows coalescing coral heads and irregular isochronous surfaces. *A. cervicornis*, not so shallow as *A. palmata* (see Chapter two, section 1.1.1), is generally the main component throughout the growth. Alacran Reef off Yucatan is an example, with a growth rate of 12 metres/1000 years relative to a sea-level rise rate of 3 to 4 metres/1000 years. Catch-up reefs occur also at Eleuthera, North Andros, in the Northern Bahamas.

The third type (Fig. 43) is the *give-up reefs*, 'relicts that founder as reef growth lags rise rate until growth declines or ceases. They deepen upward, are

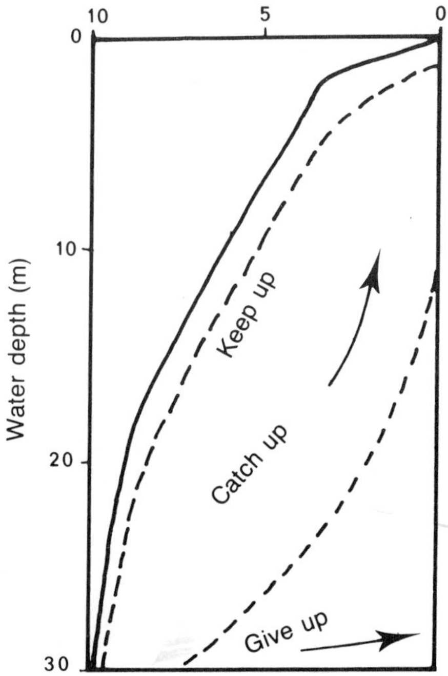

Fig. 43 Three reef responses to Holocene sea-level rise (shown by continuous line) in the Caribbean. After Neumann and MacIntyre (1985)

capped by a deep water facies, are usually small, scattered and of limited vertical extent'. These submerged relict reefs are common on insular shelves of the eastern Caribbean, e.g. the west coast of Barbados. An increase in turbidity or a cooling of shallow shelf water may have contributed to their demise. Shallow lagoons often undergo extreme conditions in temperature, salinity and turbidity, which have a deleterious influence on the reefs in front of them, so that such reefs have been said to be 'shot in the back by their own lagoons' (Neumann and MacIntyre, 1985).

Compound reefs are also found, as the reef off the south coast of St Croix (Adey, 1975; Adey and Burke, 1976), 'where an offshore catch-up reef has caught up and blocked off the ocean access of a fringing keep-up reef behind it, to the detriment of the keep-up reef'. Another example is the barrier reef off Carrie Bay Cay in Belize, where the main reef mass is of the keep-up reef type, but is now eclipsed by an offshore shelf-edge ridge which shifts towards the catch-up type with dominance of *A. cervicornis*.

A peculiarity of many Caribbean reefs, in contrast with those living in the Indian and Pacific Oceans, is the overwhelming role of massive algal ridges. This feature will be reviewed at the end of Chapter four, section 7.

In the southwest Pacific, the Great Barrier Reef has also provided many data on Holocene reef growth. Before these data are summarized, however, those on its basement, considerably improved in recent years, must be given. Previous hypothetical sections (19th and early 20th centuries) across the Great Barrier reef by Jukes, Edgeworth David, Jenson, Jardine and others, summarized in Steers (1929, p. 233–239), and Steers' own proposal (1929, p. 366), have been replaced by the result of geological and geophysical investigations (Orme and Flood, 1977; Orme *et al.*, 1978b; Symonds and Davies, 1985). In contrast to Polynesia and Micronesia (section 4 of this chapter), the history of the Great Barrier Reef has not been generally dominated by reef-building processes. After deposition of volcanic rocks during the Late Cretaceous, the facies were alternatively marine and continental, the latter, of progradational type, being dominant during Upper Oligocene, Upper Miocene and Plio-Pleistocene: they consisted of large deltaic systems (the North Queensland rivers still now provide an annual sediment input of 28 million tons). The shelf was thus widened by 10 kilometres off Cairns, and by 50 kilometres off Townsville.

Before the Holocene marine transgression, however, Pleistocene reef formations existed over this shelf. They developed during brief periods of high Pleistocene sea-level, and were dissected during the Wisconsin regression by fluvial channels, either buried or still open, which have been revealed by seismic profiling (Figs. 35 and 36). Such is the irregular basement upon which the Holocene reefs grew.

Their growth has been documented from 70 drill holes on 25 reefs in the central and northern Great Barrier Reef (Davies and Montaggioni, 1985). It occurred during a rapid sea-level rise, followed by a relative stability since 6500 BP.

In a first phase, the transgression over the colonizable substrate was dominant, and the water depth over the bottom increased. In the second phase, which began about 8000 BP, reef growth was dominant, at first with a growth rate compensating the rate of sea level change, and later with a growth rate faster than the transgression, around 6500 BP, with further leeward progradative phases. A good example is found at One Tree Reef (Marshall and Davies, 1982). If we refer to the terminology used for the Caribbean reefs, this strategy, common in the Great Barrier Reef, falls into the catch-up type, in which three subtypes have been distinguished according to relationships with the sea-level rise. The give-up type has not been identified, and the keep-up type is rare, although present at Fitzroy Reef in the southern province of the Great Barrier Reef (Davies and Montaggioni, 1985, p. 482).

In French Polynesian atolls, the basement of the Holocene corals is completely different from the Great Barrier Reef, consisting exclusively of reef formations which are reviewed in section 4 of this chapter. Holocene reef growth, known for Mururoa and Mataiva atolls (Fig. 44) and the Moorea barrier reef (Labeyrie *et al.*, 1969; Davies and Montaggioni, 1985, p. 499–503), for a sea-level history similar to that of the Great Barrier Reef,

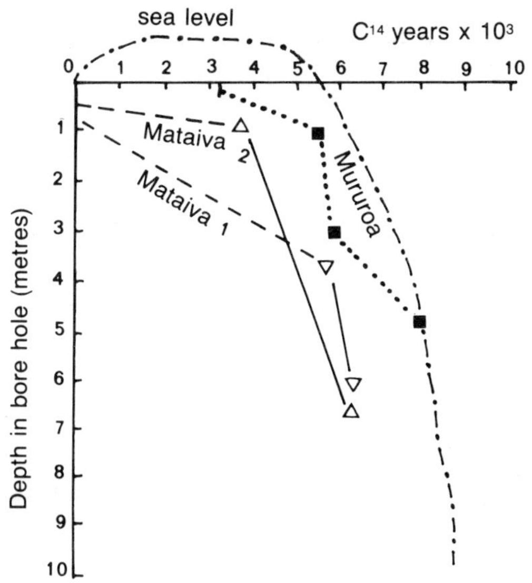

Fig. 44 Growth curves in relation to sea-level rise in Polynesia, at Mataiva Atoll (two boreholes in lagoon) and at Mururoa Atoll (Colette borehole in northern rim). After Davies and Montaggioni (convenors) (1985)

appears to have begun between 7000 and 8000 BP. Vertical accretion rates have varied between 1.5 and 2.5 metres/1000 years for the reef rim at Mururoa, and 0.2 and 2.5 metres/1000 years for the lagoon at Mataiva, with the possibility of much higher rates in places subject to hurricane sedimentation. In some places, reef growth was delayed some time after the transgression, while in others it was synchronous. The reefs of the Tuamotu and Society Islands seem thus to have followed the catch-up strategy, dependent on or triggered by the rate of sea-level rise, as in the great majority of the East Australian ones, in spite of the environmental differences.

At Tarawa Atoll, Kiribati, the Micronesian environment is not very different from that of French Polynesia, except for the lack of hurricanes. Cores have shown a reef growth beginning about 8000 BP, with fairly uniform vertical accretion at a rate of 5.0–5.4 metres/1000 years in two holes, and 8.2 metres in a third one (less accurate), with a rate of sea-level rise thought to have been 5–8 metres/1000 years since at least 12 000 BP. Tarawa thus illustrates the catch-up model too, with a growth rate higher than that of French Polynesia, and higher than at Enewetak and Bikini, Marshall Islands in Micronesia, which seem to be of similar type (Marshall and Jacobson, 1985).

These results indicate that the catch-up strategy has been extremely predominant in the Pacific, while in the Caribbean the mode of growth has been much more diversified. More data are still needed for reefs in the Indian Ocean.

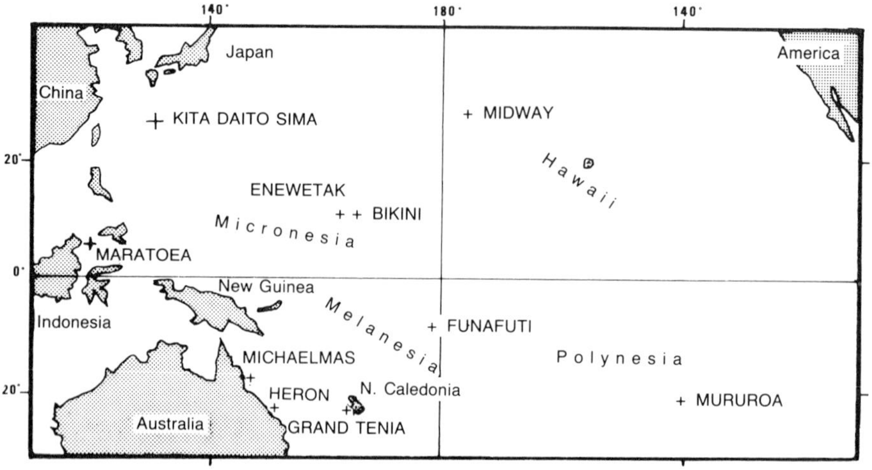

Fig. 45 Location of deep drilling in Pacific reefs, shown by crosses

4. RESULTS OF DEEP DRILLING AND GEOPHYSICAL WORK ON ATOLLS IN THE PACIFIC AND IN INDONESIA

Beside the information on shelf and reef evolution during the Quaternary in Eastern Australia and southwestern New Caledonia (Fairbridge, 1950; Hopley, 1982; Coudray, 1976), other essential data have come from deep drilling carried out in Pacific and Indonesian atolls, which began at Funafuti Atoll in the Ellice Islands, now Tuvalu (Armstrong *et al*, 1904). Atolls have been drilled (Fig. 45) at Kita Daito Zima near Okinawa, Japan (Hanzawa, 1940), Maratua off the northeast coast of Kalimantan in the Celebes Sea (Kuenen, 1947, 1950 *passim*), Bikini and Enewetak *alias* Eniwetok in the Marshall Islands (Ladd *et al*., 1948, 1960; Tracey and Ladd, 1974; Henny *et al*., 1974; Oberdorfer and Buddemeier, 1985), Midway Atoll, Hawaii (Ladd *et al*., 1970), and Mururoa Atoll, which is administratively in the Tuamotus, but belongs structurally to the Pitcairn–Gambier–Hereheretue chain (Lalou *et al*., 1966; Chauveau *et al*., 1967; Deneufbourg, 1969; Labeyrie *et al*., 1969; Buigues, 1985; Aissaoui and Purser, 1985; Gachon and Buigues, 1985; Ruzie and Gachon, 1985). The most detailed information has come from Enewetak, Mururoa, and Midway, all of them having been drilled more than once down to the pre-coralline basement, and deeply into this basement at Midway and Muroroa. At Enewetak there have been, in addition to two deep drill-holes, a number of shallower borings down to some tens of metres, but unfortunately they have all been in the atoll rim and not in the lagoon (Ladd *et al*., 1960, p. 864). At Midway one of the deep drill-holes is on the rim, and the other on an island in the lagoon (Ladd *et al*., 1970, p. A 4). Muroroa is by far the most investigated atoll in this respect since it has thirteen boreholes deeper than 175 metres, six in the lagoon and seven in different parts of the rim (Buigues,

1985, p. 251), and geophysical profiles across the atoll (Ruzie and Gachon, 1985).

All these investigations have shown (Fig. 46) the great thickness of shallow water calcareous formations in these atolls scattered around the Pacific. Two of them, Kita Daito Zima and Maratua, are raised atolls. At Kita Daito Zima the drilling stopped at 431 metres in a formation dated Upper Chattian (Oligocene). At Maratua, where two holes were drilled under the former lagoon, the deepest reached 550 metres below the crest of the rim, 429 metres below the present sea-level, also without reaching the pre-coralline basement.

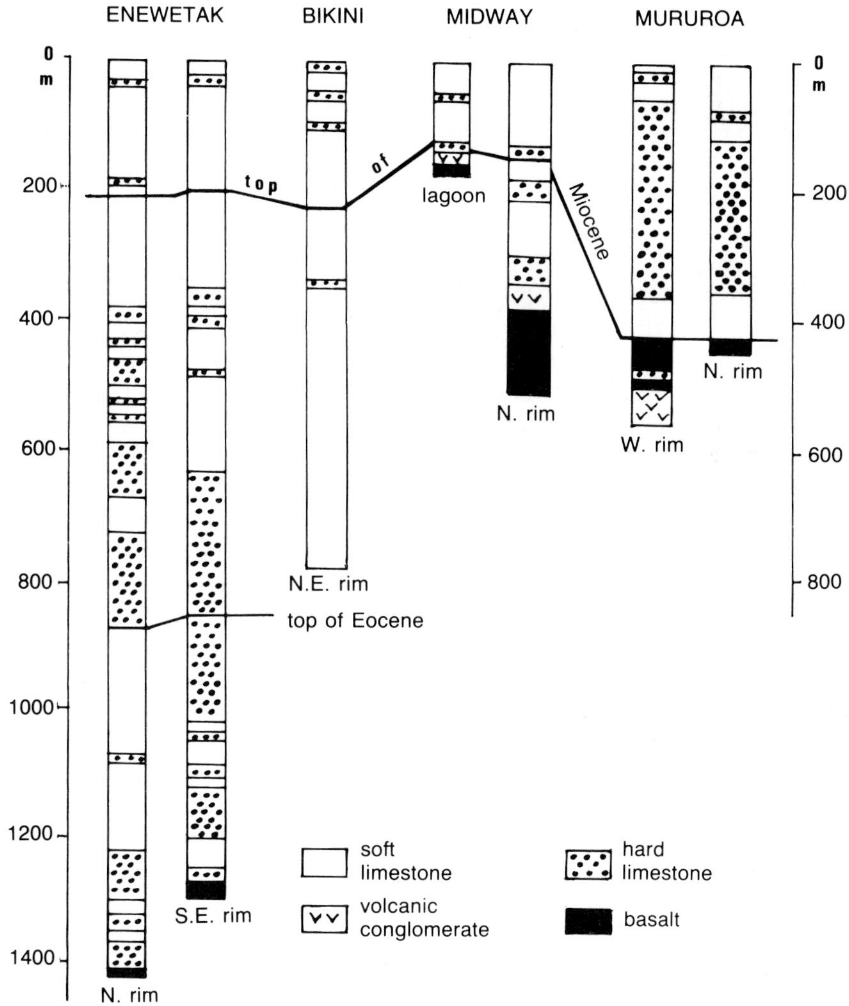

Fig. 46 Results of deep drilling at Enewetak, Bikini, Midway and Mururoa. After the US Geological Survey (1954, 1960, 1970) and Deneufbourg (1969)

The age of the rocks is not certain. At Funafuti, Pleistocene and Cenozoic rocks were encountered, but the Oligocene had not been reached at a depth of 339 metres. At Bikini, drilling to 777 metres went through Pleistocene and Cenozoic rocks without reaching the Eocene. In the two boreholes on the rim of Enewetak, the basaltic basement was reached at 1283 and 1408 metres, the oldest reef sediments, immediately above the basement, being of Eocene age, and containing corals 'that do not flourish below 150 feet' (Ladd and Schlanger, 1960, p. 899). At Midway, the basement, also basaltic, was at 173 metres in the lagoon borehole, at 503 metres in the reef rim borehole, the deepest coralline formations being earliest Miocene. At Muroroa, the two deepest boreholes found the contact between detrital coral formations and basalt at 415 and 438 metres; there the beginning of the sedimentary sequence, including the erosion of the emerged volcano, has been dated at the end of the Miocene (10 million years).

These results demonstrate, with a certainty rarely obtained in geomorphology, that the seven atolls were built by corals and associated organisms during a long-continued subsidence, over a differing period in each case. In other words, the subsidence theory, proposed by Darwin and developed by Dana, Davis and others, is thus fully supported. It remains to decide whether the atoll comes from the building, by corals and associated organisms, of a rim up to a higher level than the floor of the central lagoon (as Darwin assumed), or if this saucer shape derives from some other geomorphological process; also, whether several steps other than the classic fringing–barrier–atoll succession can be deduced from the drillings.

Before these problems are examined, evidence from the top of the basement at Mururoa and Midway (Deneufbourg, 1969; Bardintzeff *et al.*, 1985; Ladd *et al.*, 1970) will be mentioned, and also from Rurutu, a raised structure in the Austral Islands (Bardintzeff *et al.*, 1985), since they indicate how the sedimentation began. At Mururoa, five of the boreholes (Fig. 47) show a succession upwards of a submarine and subaerial volcanism now between 269 and 455 metres below sea-level, followed in four of the five by a transition level of variable thickness (up to 82 metres at Viviane), before dolomitized coralline strata are reached. The transition level consists of detrital basaltic blocks, with montmorillonitic and kaolinitic clays, overlain by calcareous rocks with variable amounts of coral. Beach sands are found, deposited in places that were shallow, while coral organisms colonized in deeper water, the pioneer species varying around the emerged volcano (Buigues, 1985).

At Rurutu, a foundation of submarine basalt is overlain by a transition zone consisting of submarine red clays and pockets of manganiferous ores, passing upwards into submarine conglomerates including a mixture of coral reef and volcanic blocks.

At Midway the basement under the reef rim consists of pure tholeiitic basalt which includes both aa and pahoehoe lavas, which have been weathered, truncated by wave action, and capped by 38 metres of volcanic clays and conglomerates with volcanic pebbles and four limestone layers.

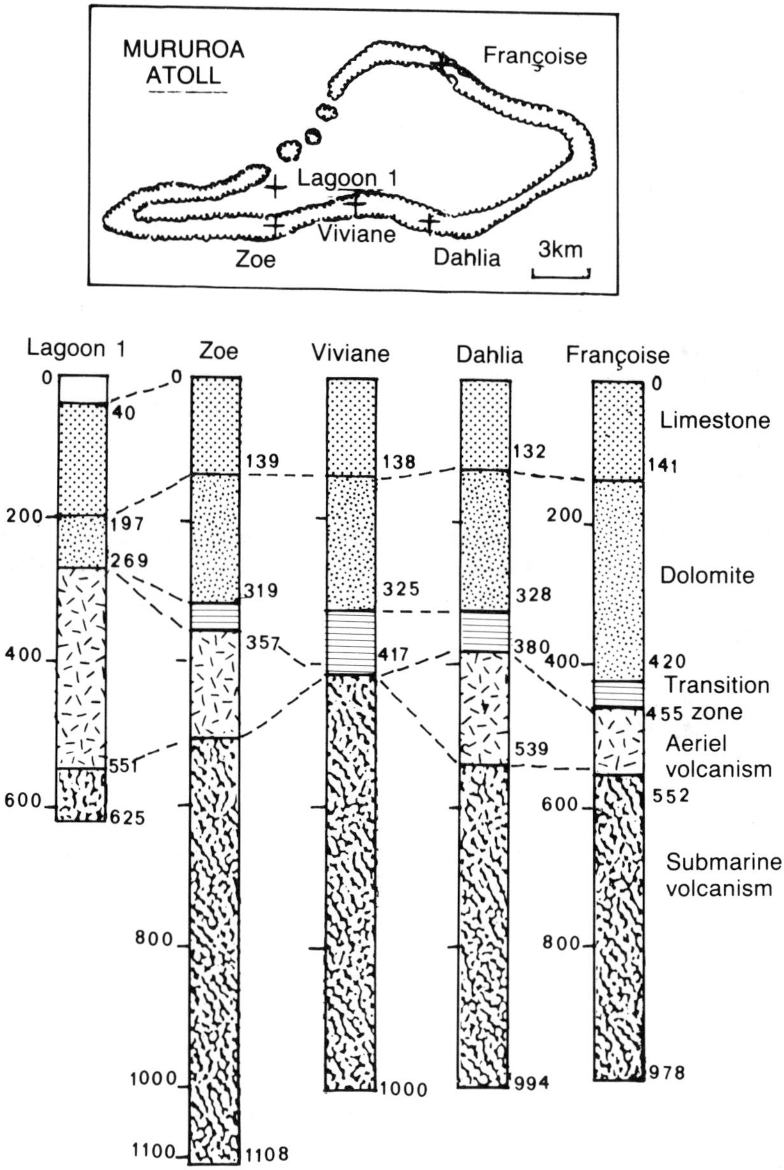

Fig. 47 Five drillings in Muroroa Atoll, showing succession of volcanism and sedimentation in different parts of the atoll. After Bardintzeff, Brousse and Gachon (1985)

All these data show that when subsidence began, the volcanoes were surrounded by beaches made of sediments derived from marine organisms and from the volcano, which was still very active at Mururoa, and in a late stage at Midway, judging by the degree of weathering and the nature of certain pebbles. At Mururoa there were successive incipient fringing reef deposits buried by lava flows during a slow subsidence, which also fit Darwin's model

Above the basal sedimentary deposits, the various boreholes show a succession, or mixture, or massive coral framestones, *in situ* algal framestones, biodetrital transported facies which are sometimes sandy, sometimes silty or muddy, and include foraminifera, molluscs, and large or small pieces of coral. The degree of cementation is highly variable, even at great depths. For example, at a depth of 1310 metres in the main borehole on Enewetak (Ladd and Schlanger, 1960, p. 894), the Eocene sediment is a 'limestone, friable to moderately well cemented, porous, white, foraminiferal–algal'. In all these drillings the internal structure of the atolls includes an enormous number of voids and caves. Several holes at Enkebi Island, Enewetak (Oberdorfer and Buddemeier, 1985, p. 310) showed in the upper 80 metres an intricate mixture of rocks described as 'boundstones', 'packstones', 'grainstones', 'wackestones', all of them either uncemented, moderately cemented, or well-cemented. French contractors in Polynesia call this kind of sediment a coral soup. The voids result in part from later solution (karstification), but some are original, deriving from cavernous reef building where the rock is not completely compacted at depth. One of the results is a freshwater lens called the Ghyben–Herzberg lens, at shallow level in the islands occurring on atoll rims (Wiens, 1962, p. 318–326), and surrounded by brackish water subject to tidal oscillations. The hydrogeological problems raised by this lens, sometimes very large as at Tarawa (Marshall and Jacobson, 1985) and extremely important in atoll life, are more complicated than it had been previously thought (Oberdorfer and Buddemeier, 1985).

In this sedimentary framework, diagenesis has occurred, leading to dolomitization which has been recognized in all the atoll boreholes, including Funafuti and Kita Daito Zima. At Mururoa, the dolomites reach a thickness of 280 metres below the peripheral islands, while they are absent from certain boreholes beneath the lagoon; they are considered to be contemporaneous with, or subsequent to, aragonite dissolution, and their distribution suggests that marine influence has been important in their formation. Internal cementation also increases towards the periphery, probably for the same reason (Aissaoui and Purser, 1985). At Midway (Ladd *et al.*, 1970, p. A 21), where the Miocene strata have been extensively dolomitized, the dolomitization is thought to have occurred during a Pleistocene submergence, following an emergence during which the limestone had been thoroughly leached and recrystallized: so that, there too, the influence of the surrounding marine environment seems to have been determinant in reef diagenesis.

In spite of the highly complicated structure of the rocks, it has been possible to find, in the upper part of the sections, general unconformities that have been identified as Pleistocene. At Enewetak and Bikini (Tracey and Ladd, 1974), 'at least four intervals represent reef growth and associated lagoonal sedimentation during relatively brief periods of Quaternary high sea levels, overlying unconformities representing periods of emergence and weathering during glacial lowering of sea level. The uncomformities separating near-reef lagoonal facies are approximately 10, 25, 50 and 85 metres below present mean low water. The equivalent reef facies for each interval would be a few metres to as much as 50 or 60 metres higher'. The uppermost marine unit is thus the Holocene one, its top slightly higher than present sea-level, as previously said. The contacts between the older units represent surfaces produced by solution during the subaerial lithification of the underlying deposits (Henny *et al.*, 1974).

At Midway, the data are not so precise as at Enewetak and Bikini, as no radiometric dates seem available; nevertheless, a large unconformity at a depth of 60 metres may correspond to the break at 85 metres at Bikini and Enewetak, and two possible higher ones may have been related to solution during later brief emergences (Ladd *et al.*, 1970 p. A 21; Tracey and Ladd, 1974, p. 547).

At Mururoa, below the Holocene reef which rises to between 1 and 4 metres above sea-level, a large unconformity caps a karstified reef, dated at ±120 000 B.P. in the Colette borehole, located in the reef rim (Lalou *et al.*, 1966). This karstified reef may thus be related to the Sangamon–Eemian–Ipswichian high interglacial sea-level. Under the lagoon, four unconformities suggest four Pleistocene sea-level variations (Buigues, 1985, p. 254).

These results fit well with Coudray's conclusions (1976, see Fig. 40, in section 3 of this chapter) from the drilling of Grand Ténia in the New Caledonian barrier reef, where he found three successive Pleistocene reefs capped by unconformities, plus the Holocene reef. In addition, the Sangamon–Eemian reef has been identified and radiometrically dated in places as widely scattered as Oahu in Hawaii; Anaa, Niau and Makatea in the Tuamotus; Mangaia in the Cook Islands; Rottnest Island in Western Australia; Mauritius; the Seychelles (Veeh, 1966); and Guadeloupe Island in the Caribbean (Battistini *et al.*, 1986).

It may be asked, however, why there are not more than three or four reefs referable to Pleistocene high sea-levels, when various studies, including Emiliani's palaeotemperature curves (1966, 1970), have indicated a much larger number of positive thermal oscillations, as well as minor fluctuations during the Wisconsin. In Alaska, Hopkins (1967) found six marine transgressions before the Holocene, including one at 2 000 000 BP, and one between 1 900 000 and 700 000 B.P. Other studies based on deep sea cores in the northeast Pacific point to glaciations (outside the Antarctic) and interglacials older than the Nebraskian/Gunz and the Aftonian/Cromerian (e.g. Kent *et al.*, 1971). It is not yet known how these variations in sea-level affected reef building, and further research is required (see section 7 of this chapter, for Papua New Guinea).

On the other hand, the numerous drillings in the Mururoa Atoll have provided a so far unique picture of atoll growth, and of variations in shape in the course of subsidence (Buigues, 1985). After the initial colonization by a fringing reef, subsidence of about 200 metres, occurring in several stages with interruptions, led to the widening of the fringing reef. It seems that a barrier reef stage, with a peripheral rim standing around a lagoon, was reached just before the top of the volcano was submerged: in other words, the barrier reef stage, which remains a little hypothetical, appears to have been short, whereas the fringing reef stage was comparatively long. Thereafter (Buigues, 1985) 'subsidence seems to have decreased and sea level variations prevailed' (as a result of early glaciations?), and the reef took the shape of 'a wide shallow lagoon extending over the whole atoll; possibly a continuous barrier reef' (actually, a rim) 'isolated this wide lagoon from the ocean'. Subsequently, the rim continued to grow vertically, while detrital, fine facies settled in the lagoon. The general difference between the rim and lagoon formations appears on Fig. 48, where borings indicate the muddy facies common in the

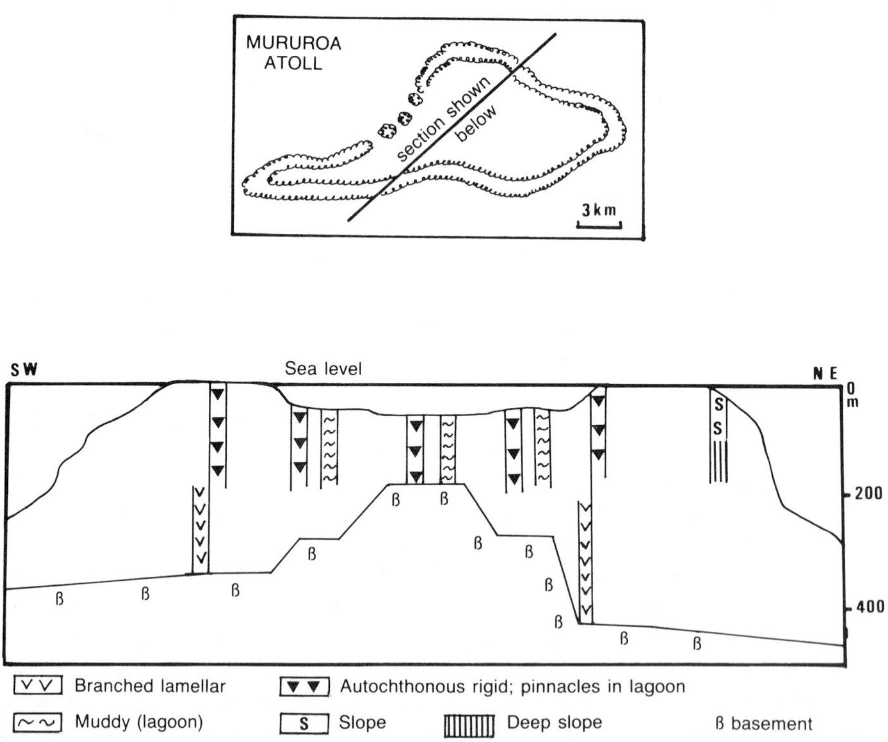

Fig. 48 Section across the rim and lagoon of Mururoa Atoll, after Buigues (1985), simplified, showing the facies at depth in atoll rim and in lagoon

lagoon, with intervening rigid frameworks representing lagoon pinnacles or ridges; while on either side the autochthonous framework or coarser biodetrital facies is predominant or exclusive, with coarse slope deposits on the northeastern outer side. The modern rim formed progressively between 90 and 40–30 metres depth.

Summarizing this history, it may be said that, in the only present-time atoll in the world in which the stages of growth shape are known with some accuracy, the fringing reef lasted longer than expected from Darwin's model; the barrier reef stage was shorter; and the atoll stage, characterized by a peripheral reef and a central lagoon with fine sediments and scattered pinnacles or ridges, much like the present lagoon, appeared when the general structure reached a level of 90 metres below the present. Apart from these slight differences, Darwin's theory is supported, especially concerning the origin of the rim surrounding the lagoon depression which lies in the atoll centre.

In the Indian Ocean, no deep drilling has been made in the open ocean reefs, except perhaps in the Maldives (see Chapter four, section 4.12). Geophysical data reported by Stoddart (1973) point to moderate thicknesses of calcareous sediments in these structures, which seem to have generally been built more recently than many of their Pacific counterparts.

5. THE KARSTIC SAUCER THEORY OF ATOLL ORIGIN

Another theory has been put forward to explain the shape of atolls by Yabe (1942), Asano (1942), Hoffmeister and Ladd (1945), Ladd and Hoffmeister (1945), McNeil (1954), Purdy (1974), Bourrouilh (1977a).

The karstic saucer theory is more or less a new version of Hoffmeister and Ladd's antecedent platform theory (1935–1944). It is assumed that it was introduced by Yabe and Asano, whose papers in Japanese were translated into English by Burke (1951) in a report of limited diffusion.

In summary, this theory states that the shape of atolls, and also of barrier reefs, derive from an antecedent, more or less horizontal, calcareous platform, which emerged as the result of uplift or sea-level fall. Having emerged, it became karstified by rainfall and percolating water, which produced the widely accepted karstic landforms such as drowned dolines and blue holes, but also resulted in a surface dissolution which gave it the shape of a saucer, with a central depression and a peripheral rim interrupted by gullies through which a part of the water escaped (the other part percolating through the karst). Subsequent submergence of such a structure resulted in revival of coral growth, which buried the karstified limestone; and the annular rampart resulting from the karstification became a barrier reef or an atoll rim. The general shape would thus result from subaerial processes and not from peripheral reef building. The explanation is not in contradiction to subsidence: it can be applied to structures as Maratua, Enewetak, Bikini and Midway, where drilling has demonstrated a great thickness of reef limestone accumulated during the slow lowering of the basement: it agrees partly with Darwin's theory since the subsidence is accepted as the cause of limestone deposition, but it disagrees on the origin of the rim.

The karstic saucer theory has found support in experiments and in geomorphological observations of emerged, atoll-shaped, reefs.

Experiments have been carried out in the laboratory on limestone blocks by Hoffmeister and Ladd (1945) and by Purdy (1947). When blocks were subjected to acid treatment, a peripheral rim was produced, standing above the centre. Purdy's results varied according to the amount of acid used: in his balance model he obtained a rim with a flat central hollow; in his undersupply model, a rim with pinnacles in the central hollow, and in his oversupply model, using an uninterrupted film of acid over all the surface of the block, neither a rim not a depression, but a general lowering of the initial platform. In the first and second models, the rim was cut by many small gullies running down the sides of the block, but it clearly stood above the central depression.

Supporting observations in nature are numerous, and scattered throughout the coral world in the Pacific, the Caribbean Sea, and the Indian Ocean.

In the Pacific, small limestone islands off the eastern coast of Okinawa, South Japan, described and illustrated by MacNeil (1954), have prominent marginal rims, which are not former built-up atoll rims. In Eua in the Tonga Islands, and in the Lau Group of the Fiji Islands, Ladd and Hoffmeister (1945) described emerged coral structures with limestone rims surrounding central depressions, which had been interpreted by earlier workers as elevated atolls. They accept that the basin shape of a part of these five islands can be interpreted as related to reef growth; but others 'do not show elevated reef sructure to support the atoll interpretation'. In other islands of the same group, the conclusion was the same. It seems probable that 'the islands were elevated from the sea as unrimmed, limestone-veneered banks', and that the saucer shape derived from karstification after uplift. At Tuvutha, also in Fiji (Fig. 49, b and c), a central volcanic core is surrounded by an inner elevated reef with ramparts, and an outer modern barrier reef. The inner ramparts are interpreted (Ladd and Hoffmeister, 1945) as karstic residual rims, and the outer barrier reef as a recent coral veneer over the edge of a calcareous Neogene platform. Also in the Pacific, Bourrouilh (1977a) has interpreted the shape of the raised atolls of Kita Daito Zima, Maré and Lifou (Loyalty Islands), Rennell (Solomons), Makatea (Tuamotus), and Nauru, as a result of karstic dissolution of the centre of a carbonate platform, and she has extended her interpretation to atolls situated at present sea-level, such as Rangiroa and Clipperton, which would represent subsequent stages in evolution (see also Chapter four, section 4.13, with Figs. 133, 134, 135 and 136).

In the Caribbean, Purdy found evidence in the large barrier reef (Fig. 49a) and three atolls of Belize. Large pinnacles in the lagoons, which he compared with the famous subaerial tropical *Turmkarst* (tower karst) described in southeastern China by Von Wissmann (1954), are thought to be derived from the drowning of similar landforms. Sparker profiles show a correspondence

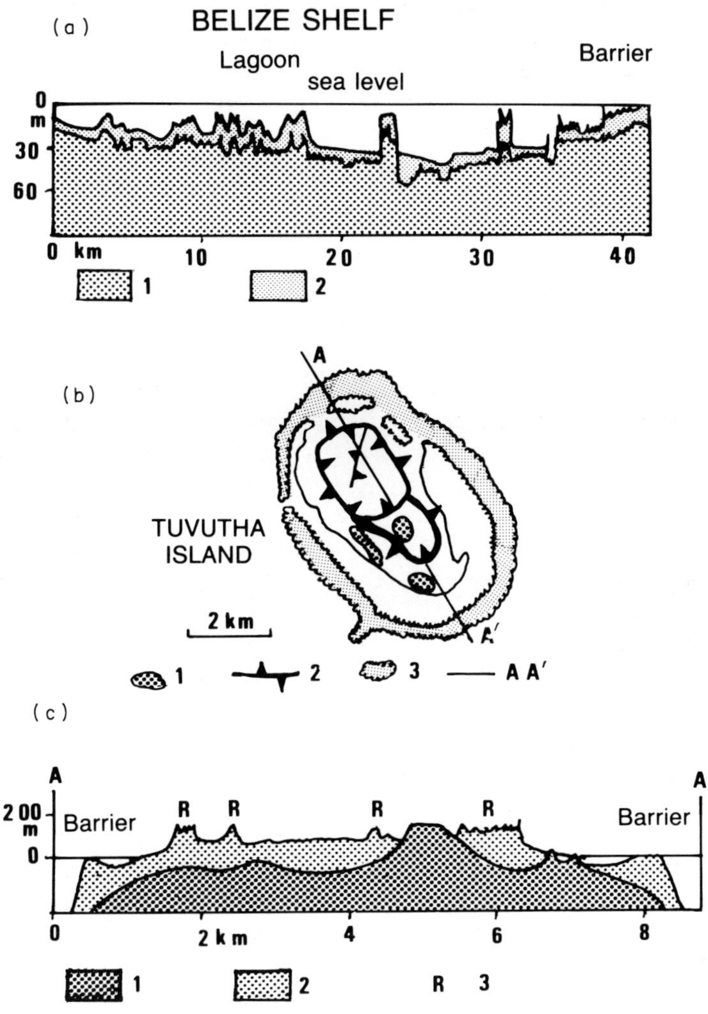

Fig. 49 Evidence from Belize and Fiji favouring Purdy's theory on karstic origin of barriers and of pinnacles in lagoons. After Purdy (1974) and Hoffmeister and Ladd (1945). (a) Sparker section across Belize lagoon and barrier. 1: Pre-Recent rocks (calcareous); 2: Holocene sediments. Shows prefiguration of present topography (barrier, pinnacles) by a pre-Holocene conical karst; (b) Map of Tuvutha Island and barrier, Lau Group, Fiji. 1: Neogene volcanics; 2: Rims in Neogene raised limestone, considered to be residual ridges produced by subaerial karstification, and not former marine rims of atolls or barriers; 3: Present barrier reef. AA′: section shown here in (c). (c) Section across Tuvutha Island: 1: Neogene volcanics; 2: Neogene limestone (and Recent in barrier and lagoon); 3: Rims (see (b) for interpretation)

Fig. 50. Longogori meandering passage across the Mayotte barrier reef, Indian Ocean, with rims on either side, perhaps remnants of Pleistocene karstification (Photo A. Guilcher)

between the pre-Holocene topography and present bathymetry. Also on the Belize shelf, calcareous ridges on either side of the English Cay Channel, a drowned pre-Holocene valley ending across the barrier reef, have been compared with the subaerial limestone narrow walls along rivers at Okinawa (MacNeil, 1954; Flint *et al.*, 1963); and similar ridges have been obtained experimentally by applying acid to a limestone block (Purdy, 1974). Counterparts of the English Cay Channel ridges exist in the Great Barrier Reef (Hopley, 1982, p. 200), and along the meandering Longogori Passage across the Mayotte barrier (Fig. 50); but there is so far no evidence of a karstic origin at Mayotte site: it is only a possibility. There is certainly a tendency for karstification, especially in the tropics, to result in long, narrow, steep-sided ridges, not only on platform sides, but also along river valleys; and such features have been reported from areas which have reefs in the vicinity. Fig. 51 summarizes the theory.

A coral structure which could be more easily explained by the karstic saucer theory than by Darwin's theory is the Chagos Bank in the Indian Ocean (Stoddart, 1971b). This coral bank is bounded by the usual outer rim, has many pinnacles in its lagoon, and depths not exceeding 100 metres; but its quite anomalous size (over 170 × 105 kilometres), completely different from ordinary atolls, especially the nearby atolls of Diego Garcia and Peros

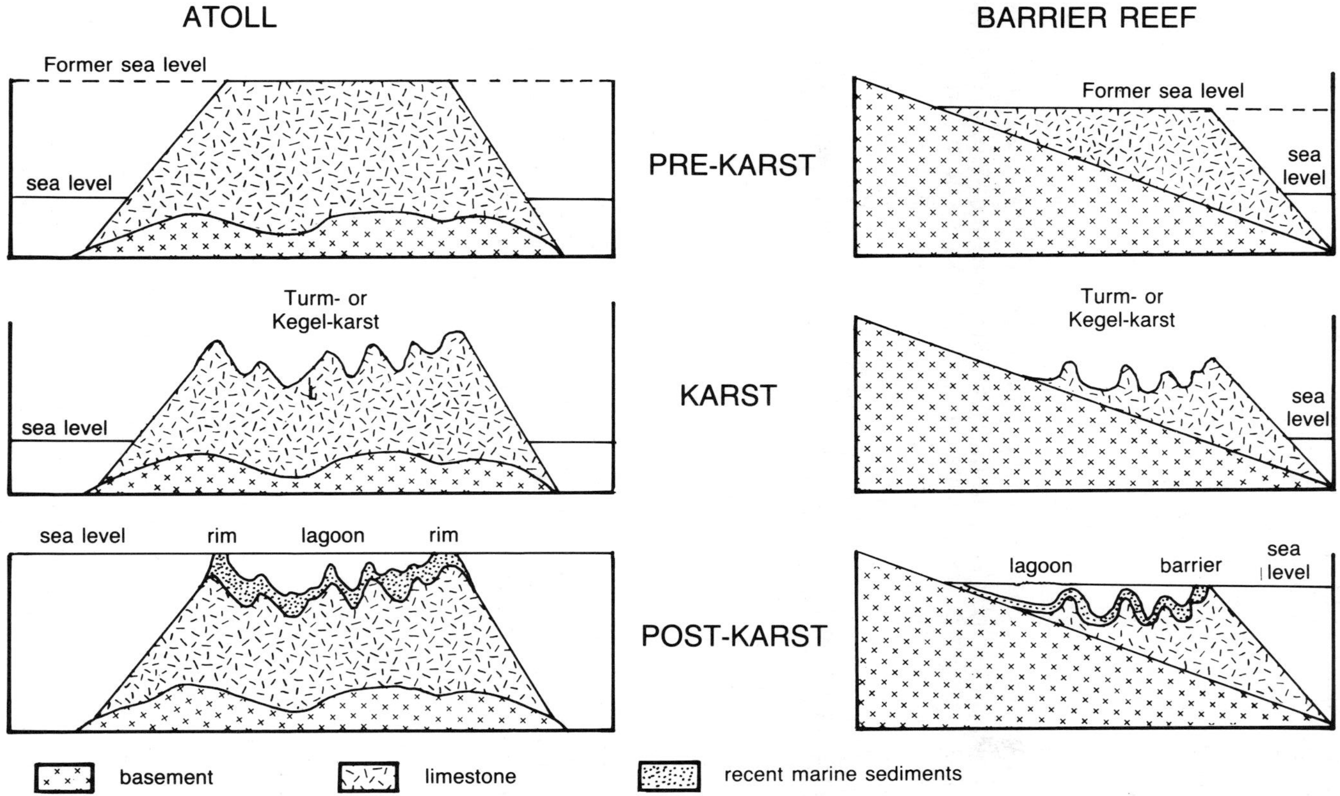

Fig. 51 Formation of barriers and atolls according to Purdy (1974) (and also MacNeil, Ladd and Hoffmeister, Bourrouilh)

Banhos, raises certainly a problem. Unfortunately the Chagos Bank is still little known, except for its bathymetry, but an explanation by the antecedent karst theory, with a reef-building starting from a calcareous platform, is a hypothesis to be checked by further research.

Purdy also offered an explanation of faros in terms of the karstic saucer theory, but this special type of reefs will be considered in Chapter four, section 4.12, for a discussion in this respect.

At the moment, which could be the conclusion about that theory in the present state of research? In 1987 we have only one set of numerous deep boreholes over the different parts of an atoll, including the lagoon: the drillings carried out at Mururoa Atoll. According to Buigues (1985), there is a pronounced difference between the framework of the rim, made of built-up reef and coarse slope debris, and the lagoon, made of fine debris except for the pinnacles and ridges; and this difference has persisted since well before the Holocene (Fig. 48). Such a structure does not favour the karstic saucer theory, but fits the classic Darwinian idea of rim gradually built around a central lagoon which has been the site of a fine sedimentation during a long period. Scott and Rotondo did not consider the Mururoa data, which were not then available, when they wrote (1983, p. 8) that 'it is clear that the present atoll morphology is fundamentally karst induced, rather than growth induced'.

Yet, this does not mean that the karstic saucer theory has no value. The evidence put forward is very extensive, especially for structures standing now well above sea-level (as those especially examined by Hoffmeister and Ladd and by Bourrouilh), but even for some of those lying at and below sea-level (especially Purdy). A worthwhile remark by MacNeil (1954, p. 415) is that 'it is necessary to interpret the history of each atoll separately instead of trying to explain all atolls by a single theory'. Beside the general fact of subsidence combined with variations in sea-level, and tectonic uplift in a number of cases, it appears wise to accept the possibility of more than one explanation of reef rim formation until many more borings have been made. It is quite necessary to drill into many pinnacles in atoll lagoons, to establish the extent to which they are Holocene structures, or Holocene coatings over Pleistocene structures. The 'Turmkarst lagoon theory', a companion to the karstic saucer theory, must be tested in a large number of sites before a general valid conclusion is possible.

6. TENTATIVE SYNTHESIS ON GENERAL REEF EVOLUTION ON THE PACIFIC PLATE

The data resulting from the deep drillings carried out in seven atolls are indispensable in any further interpretation of atoll genesis; but these data are scattered over the Pacific, while the islands in this ocean are grouped into sets or archipelagoes, generally running in a southeast and northwest direction, as in the Hawaiian, the Caroline, the Tuamotu, the Society Islands, In these groups the coral reefs and their basements vary in type from one end to the other, and belong generally to younger types in the southeast and older in the northwest. The Society Islands are typical in this respect, with a succession starting at Mehetia, on a still active volcano, passing through extinct volcanoes with barrier reefs from Tahiti to Maupiti, and ending in atolls at Mopelia, Scilly and Bellingshausen. Such a pattern has been recognized for a long time, for example by Davis who wrote (1928, p. 307): 'the eruptive construction migrated eastward with the passage of time and (the islands) have, since their eruptive construction, suffered erosion and subsidence proportional to their age'. Later Chubb (1957) commented in the same way on the other island chains of the Central Pacific. But it is only since the late nineteen sixties and seventies that development of the knowledge of deep-sea floor evolution has allowed a general explanation, ending in a genetic classification much more sophisticated than had been previously possible. Two general models have been proposed independently by Scott and Rotondo (1983a,b) and by Brousse (1985), the second one resting upon a large recent bibliography.

The sea-floor spreading theory, having been proposed separately by Dietz (1961) and by Hess (1962), and subsequently confirmed by much research, particularly the Deep Sea Drilling Project carried out by the *Glomar Challenger*, indicates that the deep sea-floor of the Central Pacific consists of a basaltic plate, which originated in the East Pacific Rise, where magma upwells to form a new lithosphere, and has drifted to the northwest. This plate collides others in the northwest and west, and disappears by subduction along trenches such as the Kuril–Kamchatka, Japan, etc. The ocean floor progressively deepens towards the subduction zone, probably because of gradual crustal cooling. Most of the atolls of the world are located on the Pacific plate, so that the consideration of its dynamics is essential for the understanding of atoll origin, development and decay, and even more the explanation and classification of coral reefs in general.

The case of the Hawaiian chain, continuing for 6340 kilometres after a bend towards Kamchatka into the submarine seamounts of the Emperor chain, can be dealt with first, since it is perhaps the simplest (Fig. 52). It is thought to have been generated by the passage of the plate over a hot spot, or melting anomaly, in the asthenosphere, located at 19°N and 155°W, resulting periodically in an active volcano. As the plate travels over the asthenosphere the successive volcanoes become extinct, are eroded, and form the foundation that has been surrounded by reefs from 24°N. In zone 2, these give way to atolls, their growth rates decreasing with increasing latitude: the decrease has been indicated by measures of carbonate production (Grigg, 1982). The contact between zone 2 and zone 3 has been called Darwin Point by Grigg, a location where corals may contribute only 20° of the calcium carbonate necessary to keep pace with recent changes in sea-level. Farther north, in zone 3, the atolls become submerged (as to the north of Kure Atoll), the peaks deepen, an accompanying deepening of the ocean floor occurs. The

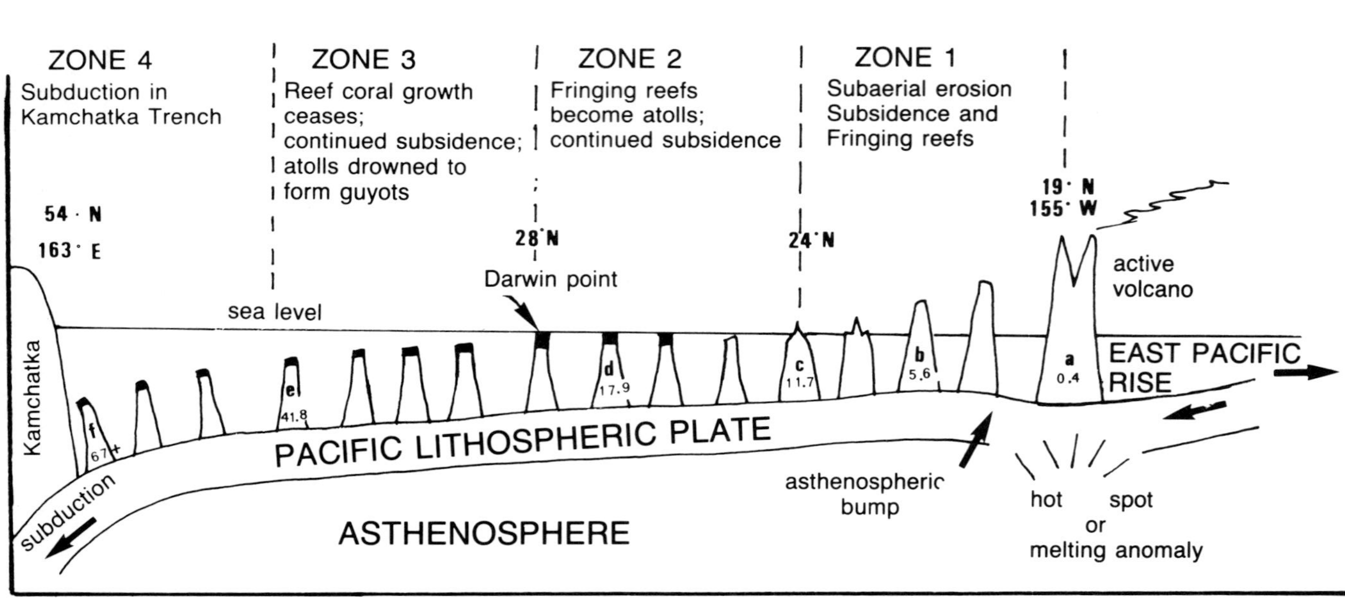

Fig. 52 Evolution of the Hawaiian–Emperor seamount chain, after Scott and Rotondo (1983), and Grigg (1982), combined. Ages of six geologic features in million years (a) Hawaii; (b) Kauai; (c) Gardner Pinnacles; (d) Midway Atoll; (e) Suiko seamount; (f) Seamount at 53°N, 165°E

passage of the Hawaiian chain into the Emperor chain is marked by a gradual bend or elbow, spread over 260 kilometres, dated at 42 million years by K/Ar and thought to result from a change in the direction of drift in the plate (Smoot, 1985). Borings on four of the Emperor seamounts have confirmed that they acquired carbonate cappings before their subsidence. The end of the story is subduction into the Kamchatka trench.

In this rather simple case there is, however, a complication related to the temporary load by the volcano built on the lithospheric plate during its passage over the hot spot. This load causes an isostatic adjustment, with a peripheral depression and a bump at a greater distance. It seems that the first author to have clearly described and analysed this moat and arch (or deep and arch) effect was Hamilton (1957) (Fig. 53), although some recent papers refer more frequently to McNutt and Menard, 1978 (e.g. Scott and Rotondo, 1983; Pirazzoli *et al*., 1985). The Hawaiian moat and arch structure is well shown on the Russian Tectonic Map of the Pacific (1966–1970) (see also Menard, 1964, p. 80–84), and Ten Brink and Watts, 1985). Similarly, the Eemian (Sangamon) coral terrace on Guadeloupe Island, Carribean, has been warped as the result of subsidence related to the formation of La Soufrière volcano (Battistini *et al*., 1986).

Data on the Society Islands chain, which is 800 kilometres long, as given by Brousse (1985), are also very accurate (Fig. 54). The hot spot is situated at

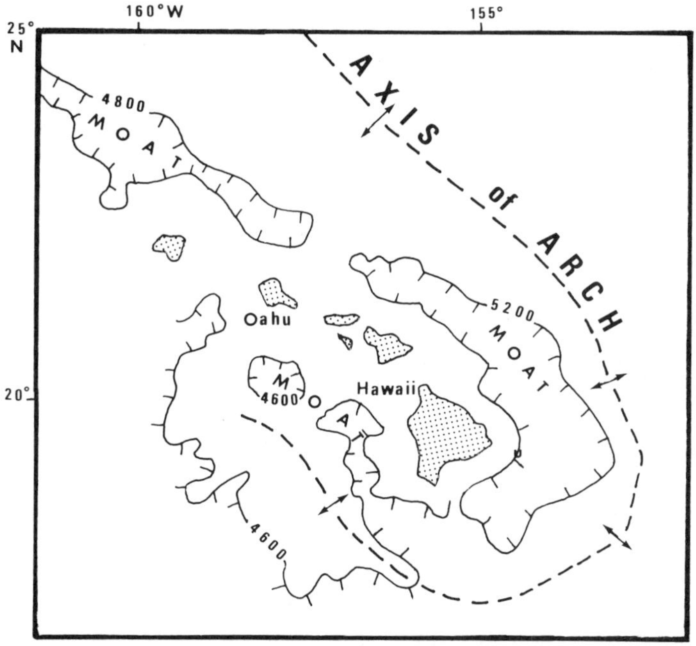

Fig. 53 The moat and arch effect around the Hawaiian Islands. After Hamilton 1957. Depths in metres

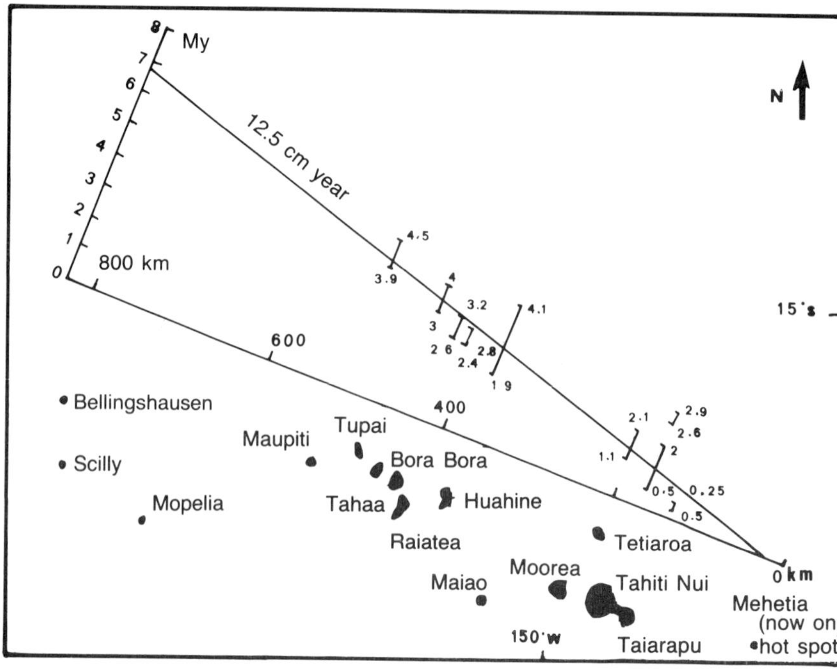

Fig. 54 Formation of the Society Islands chain, from the migration of Pacific plate over the hot spot situated where Mehetia Island is now. After Brousse (1985). Ages of volcanoes from K/Ar dating. Rate of migration: 12.5 centimetres/year. Volcanoes do not rise above sea-level west of Maupiti

Mehetia Island, a small active volcano, of small size since it is not fully developed or because of a decrease in hot spot activity. From Mehetia to Maupiti, the last volcano still visible at the surface, and surrounded by a barrier reef, eight K/Ar ages have been obtained, which indicate a rate of propagation of 12 centimetres/year.

The Pitcairn–Hereheretue chain, 1800 kilometres long (Fig. 55), through Gambier, Fangataufa, Mururoa and Duke of Gloucester (Brousse, 1985), disappears in the northwest at the southern end of the Tuamotus properly. This chain consists of two volcanic islands, Pitcairn, and Gambier which is an almost-atoll with small high islands such as Mangareva (Fig. 95); all the others are true atolls. The chain is of the same type as at Hawaii and in the Society Islands; the rate of propagation, evaluated from K/Ar measurements at four sites, is 10.7 centimetres/year, i.e. not very different from the Society Islands. The age of the westernmost and oldest atoll, Hereheretue, is 15 million years.

The Tuamotu Islands *sensu stricto* are 2000 kilometres long, and also run SE–NW, but they include only atolls, and have no volcanic outcrop. No drilling has yet been reported, but some information is available from the DSDP (*Glomar Challenger*) drilling in the sea-floor around their northwest part.

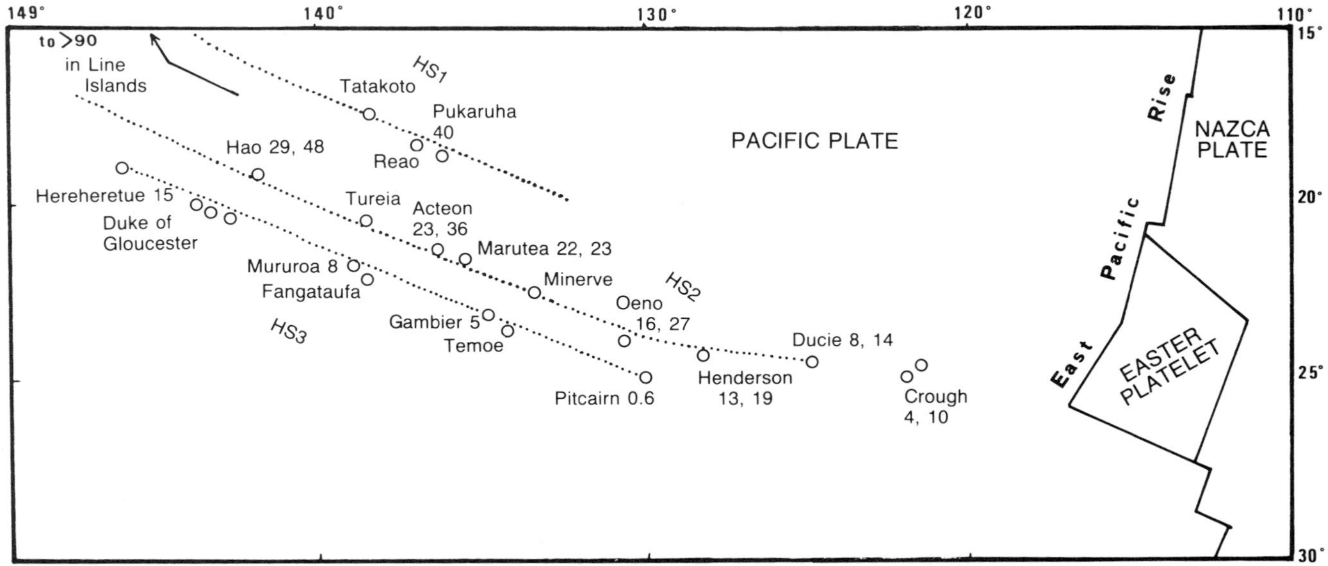

Fig. 55 Assumed formation of atolls in the Southeast Pacific by drifting of the Pacific plate over three hot spots (HS1, 2, and 3). Solid lines: plate boundaries. Lines of dots: tracks after passage over hot spots. Circles: atolls, high islands and seamounts. Numbers indicate the age of the structures in Ma (second number along HS2: age of plate on which structure sits). Eastern part of HS2 surficial expression is trapped by a fracture zone, hence the deviation of the chain. After Okal and Cazenave (1985) for HS1 and HS2 chains; after Brousse (1985) for HS3 chain

They continue, after a bend similar to the elbow between the Hawaiian and Emperor chains, and probably related to the same event in plate spreading, by the Line Islands chain, also atolls and some seamounts. The interpretation of the Tuamotus and the Line Islands remains controversial (Okal and Cazenave, 1985; Brousse, 1985), but they are certainly among the oldest structures bearing atolls in the Pacific. If the hot spot concept is applied (Fig. 55), at least two spots are required, and perhaps more. The ages proposed (Fig. 55) are only tentative, but in any case they increase from southeast to northwest and reach 92 million years in the northern atolls of the Line Islands (Schlanger *et al.*, 1984): in other words, atolls such as Palmyra and Washington began to grow during the Turonian. Near Rangiroa in the northwestern Tuamotus, a *Glomar Challenger* boring has found volcanogenic siltstones containing redeposited reef fossils dated for a part of them at 51 million years (Lower Eocene). At the other end of the Tuamotus, the eastern part of the Hao–Crough chain (HS2 on Fig. 55) shows a bend which is considered to be the expression of a fracture zone deviating the signature of the hot spot. Henderson Atoll has been uplifted at 30 metres (Fosberg *et al.*, 1983), an uplift related, according to McNutt and Menard (1978) and Okal and Cazenave (1985), to the moat and arch effect of the recent island of Pitcairn in the HS3 chain.

Similarly, in their northwestern section, the Tuamotu Islands are not far from Tahiti, the last but one volcanic island of the Society chain. According to Pirazzoli and Montaggioni (1985), this is another example of the moat and arch effect. The weight of Tahiti seems to have produced a rise of the southern parts of the atolls of Tikehau, Rangiroa, Kaukura in the Tuamotus (Fig. 34), resulting in an anomalous height of Pleistocene reefs of these islands: the jagged *feos* (Fig. 56), which are unknown as such elsewhere in the Tuamotus. The raised atoll of Makatea (Fig. 34) would have also been influenced by the rise of the arch, but its altitude (113 metres) cannot be fully explained in this way (see Chapter four, section 4.13). The westernmost atoll of the Tuamotu chain, Mataiva, is a very unusual structure which will be discussed in Chapter four, section 4.8.

The Austral–Cook Islands chain (Brousse, 1985) is 2500 kilometres long. It begins in the southeast at the MacDonald active submarine volcano, and the propagation rate has been calculated from K/Ar dating at 11.4 ± 2.3 centimetres/year. Mangaia in the west is 19 million years old. But, after a normal sequence from fringing reef to barrier reef, a volcanic reactivation has occurred in several islands (Rurutu, Rimatara, Mangaia, and others), so that the reefs of these islands have been raised 70–100 metres. One explanation, put forward elsewhere by Scott and Rotondo (1983), would be that these volcanoes, after their initial formation over a hot spot (where MacDonald is now), have passed over an asthenosphere bump. Although this explanation is not certain, it should probably be taken into account for several isolated atolls in the Pacific.

The Caroline Islands chain (Scott and Rotondo, 1983) is a rather simple

Fig. 56 Rangiroa Atoll, Tuamotu Islands. Feos (emerged corals dissected into jagged forms by bioerosion) standing on the south rim of the atoll. Height: 2.50 metres. Age: Pleistocene (not precisely known) (Photo A. Guilcher)

case, like the Hawaiian–Emperor chain, with volcanic islands in the south-west, the almost-atoll of Truk in the middle, atolls in the west, and final engulfment into the Mariana trench. There are only some complications at the western end; and, as this chain does not reach higher latitudes when it enters deeper water, the problems of coral growth found on the Hawaii-Emperor chain do not occur.

The Kiribati (Gilbert)–Marshall group is more difficult to classify. It lacks high volcanic islands in its southeastern end. The many, fine atolls of the Marshall Islands are scattered on two main lines, and are associated with numerous seamounts; some of them are flat-topped, having reached the surface (the so-called guyots), others not. In the northwest this group shows the usual deepening of the sea-floor, and includes only submarine seamounts. It is divided (Fryer and Smoot, 1985) into the Dutton and Michelson ridges, and is subducted into the Mariana and Izu–Bonin trenches. Seamounts smaller than 40 kilometres in diameter have been broken up by the fracturing of the plate surface as it bends prior to subduction, whereas the larger ones seem to be subducted intact. The Kiribati–Marshall group represents probably, like the Tuamotus, a senile chain produced by a long-inoperative hot spot. The Eocene age of the basement in the Enewetak drilling, which is still far from the western end, is evidence for such senility.

Fig. 57 Upolu Island, Western Samoa, southwest end. An example of a wide fringing reef (Photo A. Guilcher)

Fig. 58 Rose Atoll, American Samoa, the only atoll in the Samoa group, with one passage on leeward (western) side and one islet on windward side (Photo A. Guilcher)

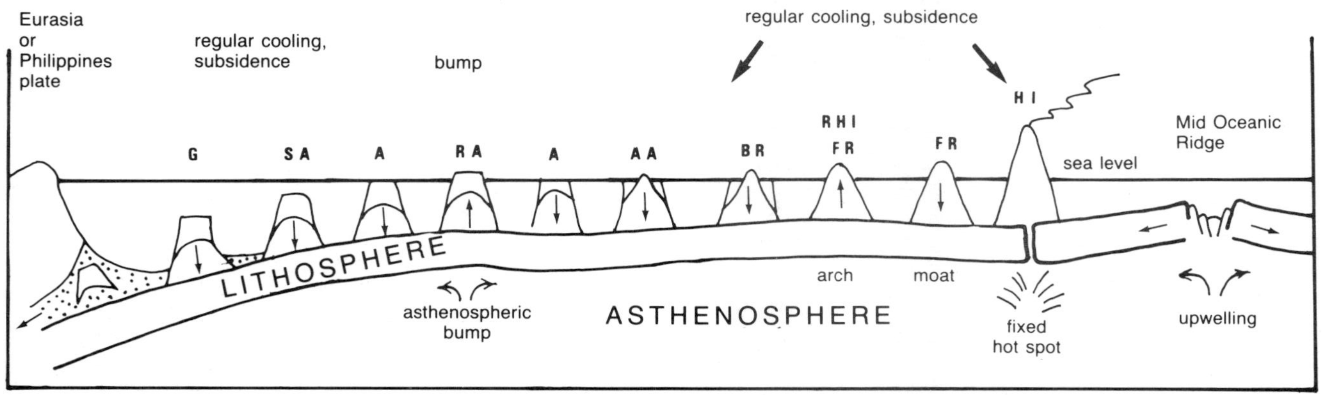

Fig. 59 Proposed model for island development on the Pacific plate. Simplified after Scott and Rotondo (1983). HI: high island; FR: fringing reef; RHI—FR: raised high island with fringing reef; BR: barrier reef; AA: almost-atoll; A: atoll; RA: raised atoll; SA: submerged atoll; G: guyot. Arrows indicate subsidence or emergence

The Samoa chain is exceptional in that it includes active volcanoes with fringing reefs in the west (Fig. 57), and only one atoll, Rose (Fig. 58) and seamounts in the east: in other words, the trend is reversed here. This anomaly might be related to the location of Samoa just at the border of the Pacific plate, i.e. close to Melanesia along the Tonga trench; in Melanesia, the polarity of the trenches has been sometimes reversed in the course of time, as in the Vanuatu (New Hebrides) arc system in the Late Tertiary (Karig and Mammerickx, 1972). The exception could be that this chain, owing to its marginal position, has not followed the general rules of the Pacific plate, but this suggestion needs to be checked.

A general model is proposed for the Pacific plate, the Samoa chain being left out, after Scott and Rotondo (1983a) with some simplifications (Fig. 59). The moat and arch effect modifies initially the subsidence which continues subsequently. The passage over an asthenospheric bump results in a raised atoll, which will sink later, leading to submerged atolls and seamounts (guyots). It may be that the moat and arch effect acts also at the passage over a bump, giving a subtype of raised atoll, supposed by Scott and Rotondo but not shown here. Some atolls (principally the Phoenix and the Tuvalu) do not fit this model: they will require further explanation, but most of the coral structures found on the Pacific plate are satisfactorily explained in this way. The reefs of the Western Pacific and Southeast Asian seas, outside the Pacific plate, where tectonic effects are essential, are examined in section 7 of this chapter.

A problem is whether the Pacific plate coral reef model finds counterparts outside that area, particularly in the Indian Ocean. The Comores Islands and their eastern surroundings, lying at the northwest of Madagascar (Fig. 60), seem rather favourable at first (Guilcher *et al.*, 1965a; Esson *et al.*, 1970; Battistini and Cremers, 1972; Hajash and Armstrong, 1972; Nougier *et al.*, 1979; Maugé *et al.*, 1982). The Comores Islands proper comprise, from west to east, the Great Comore, with an active volcano and discontinuous, incipient, fringing reefs; and Moheli and Anjouan Islands, two extinct volcanoes, 400 00 to 1 500 000 years old, with well-developed fringing reefs. Then comes Mayotte Island, an extinct volcano 3 to 4 million years old, with a fine double barrier reef surrounding a wide lagoon, and a recent (perhaps only a few thousand of years old) revival of eruptions, which have built a small volcano across the eastern barrier since its formation (Fig. 85). Still farther east are found three seamounts rising from the deep-sea floor: Zélée Bank, a slightly drowned atoll (rim depth between 10 and 17 metres, and a former lagoon at 40 metres); Geyser Bank, topped by a crescentic reef open to the east, a half atoll or tilted atoll (Fig. 127); and the Glorieuses Islands, a slightly emerged, Eemian/Sangamon reef. The cores of Zélée, Geyser and Glorieuses are volcanic. The resemblance with chains on the Pacific plate is striking, the general pattern suggesting the passage of a plate travelling from west to east over a hot spot now located at the Great Comore, the influence of a further bump at Mayotte, and perhaps another one at the

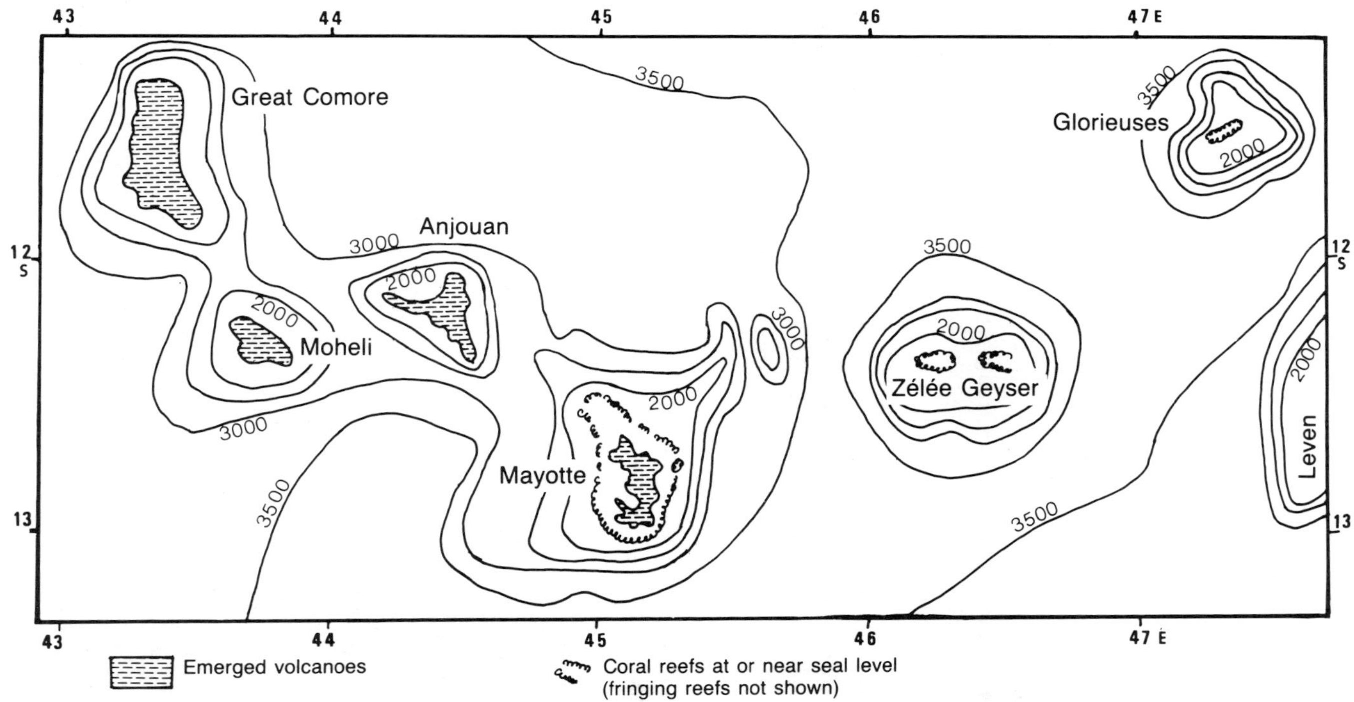

Fig. 60 The Comores chain, with the 3500, 3000, 2500, 2000 isobaths. Leven Bank is attached to the Madagascar shelf

Glorieuses since they stand a little higher than Zélée and Geyser. Unfortunately, recent data on that area have not provided evidence for such a sequence. The nature of the crust beneath these islands and banks is assumed to be oceanic, but this is still a hypothesis (Scrutton *et al*., 1981). In the present state of knowledge (1987), the mechanism of the formation of this set of volcanoes and accompanying coral structures remains uncertain.

7. IMPACTS OF REGIONAL TECTONICS

Major objections to Darwin's theory, based on the study of uplifted coral structures, have come mostly from the Western Pacific, outside the Pacific plate (they are not really valid since, as we have seen, Darwin himself had left room for such cases). These uplifted frameworks occur also in Indonesia between the Pacific and Indian Oceans, in Melanesia, in the Japanese archipelago, and in the Red Sea; they deserve attention here as they have resulted in interesting landforms (Figs. 61 and 62). Other examples will be found in Chapter 4, sections 3.5, 4.13 and 5.

The Japanese Islands lie close to the trench into which the Pacific plate is engulfed, so that the coral reefs growing on their eastern margin are subject to uplift when phases of engulfment occur. The southern part of the Boso Peninsula near Tokyo bears Holocene fringing reefs dated at 6500 BP, and

Fig. 61 Vate Island, Vanuatu, Melanesia. Two uplifted coral terraces in Vila Bay (Photo A. Guilcher)

Fig. 62 Erromango Island, Vanuatu. Cook Bay. A coral terrace, 3–4 metres high, jagged by bioerosion, standing above the present corrosion platform. Corals grow now in front (Photo A. Guilcher)

located in embayments at Numa, Tateyama City, which have already been uplifted to a height of 12–18 metres (Matsushima, 1979; Nakata *et al.*, 1979; Yonekura, 1975).

This is still a simple process, not leading to spectacular morphological features; but a more imposing result of a strong and long-continued uplift is a *set of theatre-like terraces*, the most impressive being on the north coast of the Huon Peninsula, Papua New Guinea, facing New Britain (Chappell, 1974). In that area, recent and still active uplift of the peninsula has resulted in more than twenty successive fringing or barrier reef terraces, partly associated with deltas, and ranging up to more than 600 metres high and 80 kilometres long. They stand like a gigantic Greek amphitheatre, well visible because the area is treeless and covered with grass. Tectonic uplift has varied along the terraces, so that their heights diminish from southeast to northwest, parallel to the coast. By means of radiometric dating of the terrace reefs, a record of sea-level changes relative to the rising land has been obtained. Such a large and rapid uplift has resulted in a separation of the successive high sea-level stillstands which can hardly be perceived in more stable areas. Eight sea-levels have thus been recognized and dated from 30 000 to 220 000 BP, in good agreement with what has been described in the Caribbean, the Mediterranean, Japan, and California. For the last 400 000 years, an agreement has been found with Emiliani's (1966) temperature curve. These results must be compared with those derived from the drilling of atolls and the New Caledonian barrier reef, where the relatively small number of Pleistocene high sea-levels has been noticed (this chapter, sections 2 and 4).

The raised coral reefs of the Huon Peninsula have also given an opportunity to study relations with deltas and to define littoral, lagoonal and reef environments. Notches in uplifted reefs indicate short stillstands, and were cut when the next fringing reef was already being formed at a lower level. An example of a reef intermediate between fringing and barrier is described in Chapter four, section 1.5 (Fig. 76).

Although beautiful, the Huon set of coral terraces is by no means unique, and falls into a special form of coral reefs. A description of such terraces was given by Schick (1958a,b) from the Tiran Island at the entrance of the Gulf of Elat or Aqaba in the northern Red Sea (Fig. 63). This set rises to 320 metres and includes at least eleven terraces, with more doubtful relics higher up. In Melanesia, other sets of raised reef terraces are known, and have been described from several islands (Efate, Malekula, Santo) of Vanuatu (New Hebrides). They rise to more than 300 metres in some places (Obellianne, 1958; Mitchell, 1968; Guilcher, 1974; Lecolle and Bernat, 1985). Counterparts were discovered in Indonesia at Timor and Atauro Islands (Chappell and Veeh, 1978).

At Guadalcanal in the Solomon Islands (Guilcher, 1965) (Fig. 64), a

Fig. 63 The western series of coal terraces at Tiran, Red Sea (Photo A. P. Schick) (Reproduced from *Geogr. Ann.* **40** (1958), with permission.)

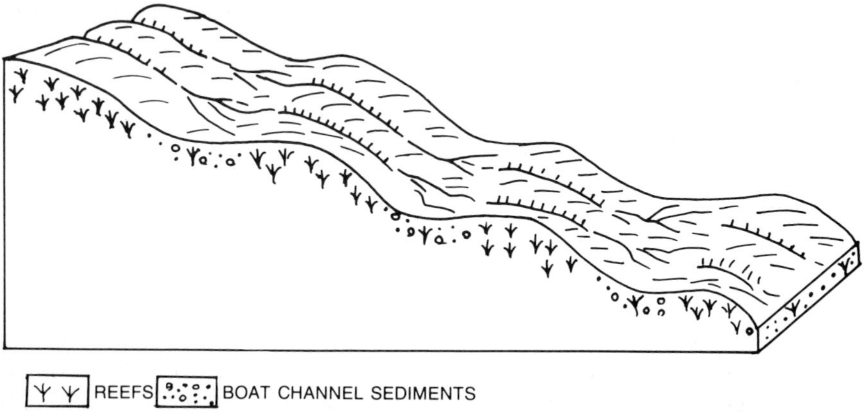

Fig. 64 Reef staircase at Honiara, Guadalcanal, Solomon Islands. From Guilcher (1965)

similar reef staircase exists in the vicinity of Honiara, rising to 210 and perhaps 270 metres, and resulting from a combination of eustatic shifts of sea level, general tectonic uplift, and a moderate tilting. When each of the fringing reefs emerged, another one was initiated at a lower level, with a shallow boat channel in between. The terraces are treeless, as on the Huon Peninsula, and thus quite visible. Sections show that the boat channels periodically received, during floods, a terrigenous sedimentation. Alternating with coral growth, such successions are similar to the deltaic deposits reported at Huon by Chappell, and testify to the wet climatic environment in which these two staircases were formed.

Another very fine set of coral terraces exists at Hispaniola Island, in the northwestern part of Haiti, but it has not yet been investigated in detail in the field (general setting shown in Vila *et al.*, 1986).

Perhaps the most fantastic example of tectonically-influenced reefs is the elevated double barrier reef of Marovo, at the eastern end of the New Georgia group in the Solomon Islands, described by Stoddart (1969c,d). This double barrier reef, facing east, consists of two lines of coral islands surrounding the volcanic (mainly andesitic) island of Vangunu, and, to the south, the other volcanic island of Gatukai (Figs. 65 and 30). The inner barrier is separated from the outer by water more than 180 metres deep. The outer barrier rises from 13 to 25 metres above the datum, the inner one being from 0 to 7.5 metres high. This double barrier, which passes into a single one in the north and in the southwest, surrounds a lagoon very irregular in depth, with many basins and steep-sided shoals, depths of more than 55 metres having been charted. The outer barrier islands have generally vertical or overhanging cliffs on their seaward side, and are deeply incised by elevated intertidal corrosion notches. The largest and most extensive notch, which was measured is at +4 metres, is floored by a layer of algal limestone, and has a maximum

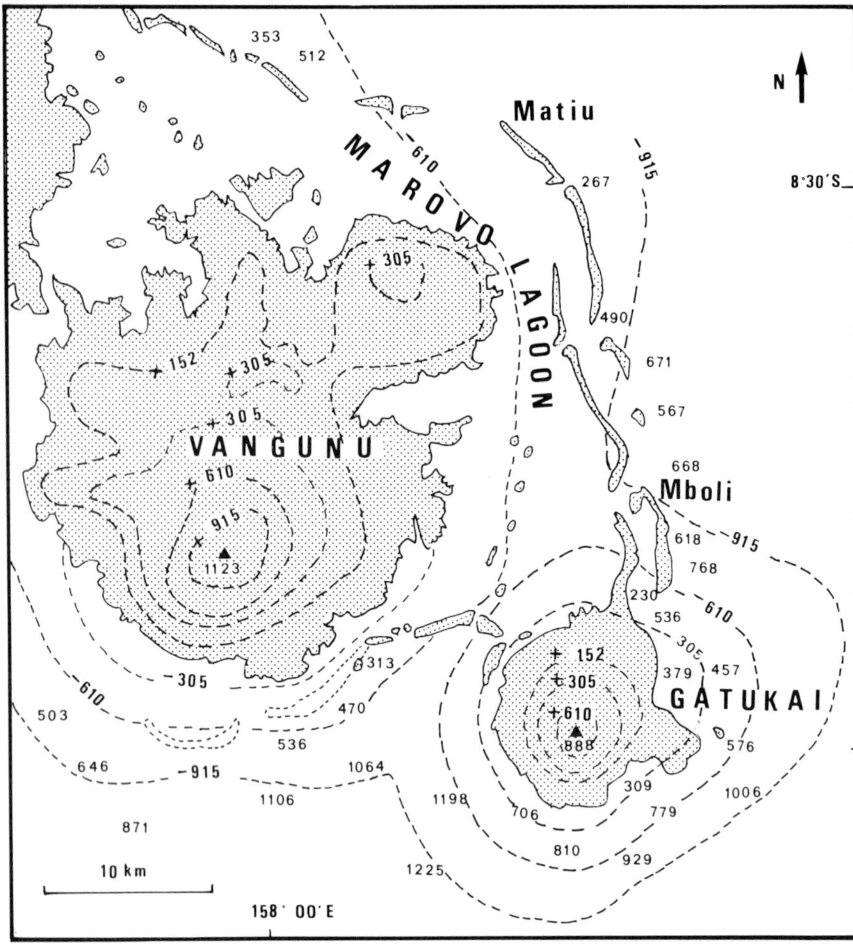

Fig. 65 Topography and projected underwater and sublevel contours on Vanganu and Gatukai volcanoes, New Georgia Group, Solomon Islands, with the elevated double Marovo barrier reef. Altitudes and depths in metres. After Stoddart (1969).

depth of 15 metres; its visor or upper rim is a complex feature with two or more concave sections; slatactites and stalagmites occur in the notch, just as in a subterranean cave. A smaller notch exists at a lower level, and submerged terraces have been reported at depths of 10, 15 and 45 metres below sea-level. From an analysis of the site and projection of underwater and subreef contours of the Vangunu and Gatukai volcanoes, Stoddart (1969c) estimated that the Marovo reef prism was 'not less than 300 m and possibly more than 600 m thick at its outer edge'. The most probable sequence of events is that it was initiated in the Late Tertiary on the subsiding volcanoes 'in the Darwinian manner' (as Davis already deduced in 1928, p. 399, from much poorer data than Stoddart); the barrier reef then resulted from further subsidence. This occurred in stages where the barrier is double, and was more uniform where the barrier is a single formation (see Chapter four, section 3.5, for double and multiple barriers in general). This barrier was subsequently raised tectonically in several phases; emergence of the 4 metres notch has been ascribed to tectonic rather than eustatic changes. The irregular topography of the lagoon seems to be of karstic origin, developed during Pleistocene low sea-level(s). Modern coral growth is very moderate. What remains to explain is the very large depth—more than 180 metres—of the channel between the two barriers; it may be due to local downfaulting. The Marovo

Fig. 66 Northwestern coast of Vella Lavella Island, New Georgia Group, Solomon Islands. The fringing reef in the foreground passes into a barrier reef in the distance as the result of tilting (Photo A. Guilcher)

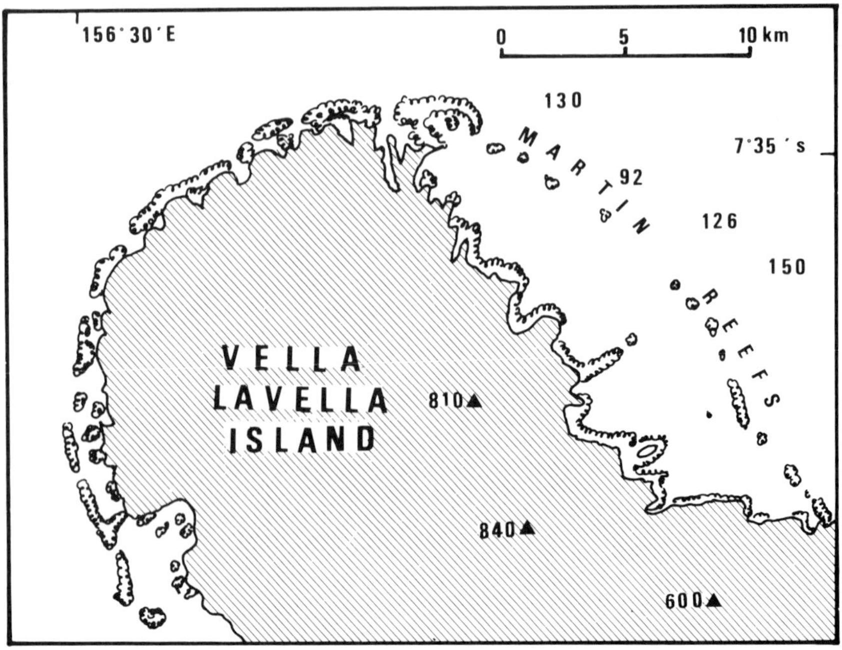

Fig. 67 Reefs around the northern part of Vella Lavella Island. Data from Admiralty chart 3419. Heights and depths in metres

reef is thus a fine example of the combined effects of subsidence, tectonic deformation and eustatic shifts in sea-level.

Variation in tectonic influences from place to place in the New Georgia group is obvious, for example, on the north coast of Vella Lavella Island (Figs. 66 and 67) where a fringing reef passes laterally on both sides into a barrier reef, with a progressive development of lagoons between the reef and the mainland. This is probably a result of local subsidence of the mainland, and is a striking illustration of Darwin's theory. Similar features exist at the eastern end of Guadalcanal.

In all the tectonic influences so far analysed, no evidence against Darwin's theory has thus been found to explain thick reefs. Only thin reefs, as those forming theatre-like coral terraces, escape this theory. On the other hand, the karstic saucer theory apparently does not find a support in the Marovo barrier reef and lagoon, since the supposed karstic features in this lagoon are not an essential fact. However, raised reefs such as those described in the Fiji Islands (Fig. 49) seem to exemplify combinations between the shaping of large internal depressions in reefs by subaerial solution and the effects of tectonic uplift.

As a general conclusion on reef genesis, Steers and Stoddart's statement at the end of their 1977 review may be reminded as being always quite acceptable: Darwin's subsidence theory provides generally adequate explanation of the development of the main reef types; Daly's theory in itself is inadequate, but Quaternary changes of sea-level have been of crucial importance, particularly in permitting the karstification of emerged reefs. Regional tectonic upheavals explain certain reef features in marginal regions such as Melanesia and the Red Sea, as Darwin noted. But if extensive drilling has greatly clarified the evolution of Central Pacific reefs, similar boreholes are needed in the Indian Ocean, the Red Sea and the Caribbean (outside the Bahamas, already investigated: Chapter four, section 6.1) to improve our knowledge of reef genesis.

CHAPTER FOUR

Types of reefs

The traditional division of reefs into three main types is retained in this book. Fringing reefs, barrier reefs, and atolls will thus be examined successively, with examples and variations of each type. However, bank reefs and patch reefs are sometimes considered as a special type, and will be dealt with here just after fringing reefs. Ridge reefs, so far neglected, will be introduced: they occur mainly but perhaps not exclusively in the Red Sea. Another section will deal with special cases which do not fit this classification. Finally, the contrasts between Indo-Pacific and Atlantic reefs, including those of Brazil, will be reviewed.

1. FRINGING REEFS

1.1. Fringing reefs without boat channel

It is unusual to find fringing reefs without any boat channel, since there is generally some sediment arriving from the adjacent land which prevents good coral growth close to the shore. Such fringing reefs are found, however, particularly on the Red Sea coasts where, owing to the arid climate, the supply of sediment is frequently meagre. A good example, described with precision by Mergner and Schuhmacher (1974), occurs 6 kilometres south of Aqaba harbour in Jordan (Fig. 68a).

This coastal reef has been built off a narrow beach of sand and pebbles, with rock slabs. The reef flat is 60 metres wide, with rich coral growth underlain by reef rock without large voids. The reef flat is about 0.80 metres

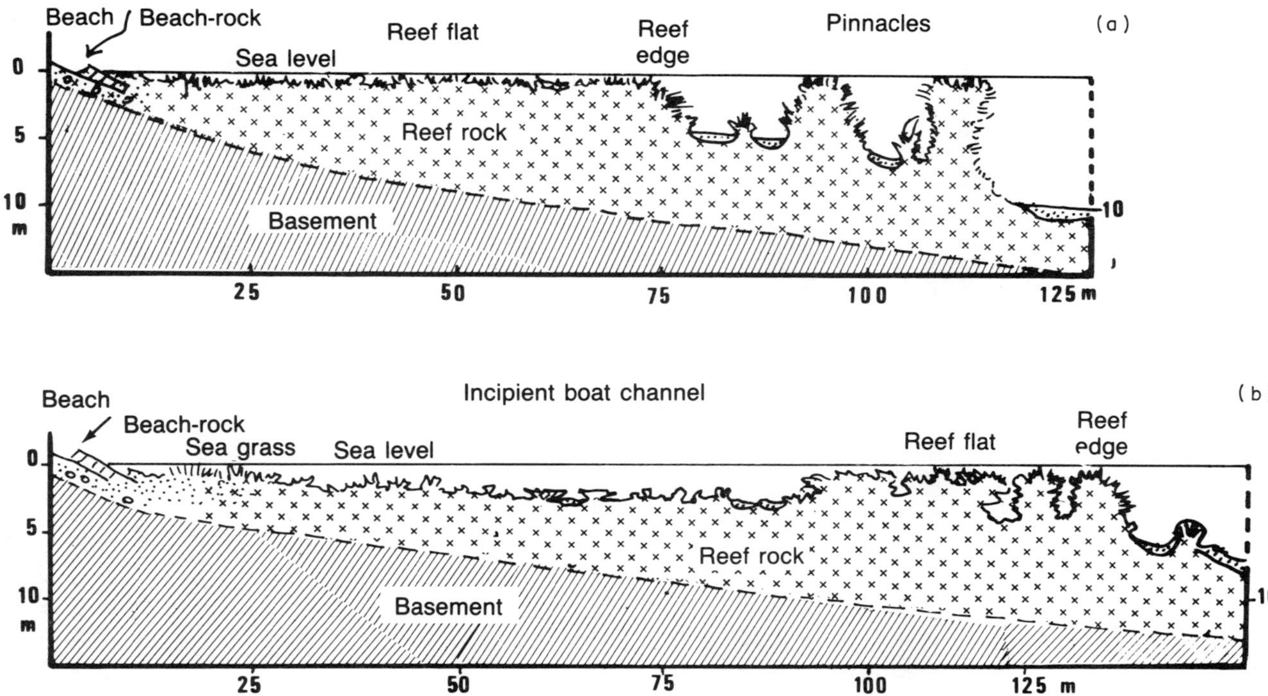

Fig. 68 (a) Fringing reef without a boat channel, 6 kilometres south of Aqaba harbour, Jordan, Red Sea; (b) Fringing reef with an incipient boat channel, 6.75 kilometres south of Aqaba harbour, Jordan. Both figures simplified from Mergner and Schuhmacher (1974)

below sea level (tidal range is insignificant, but the level can be lowered by meteorological and seasonal variations). The outer edge is normally steep, with horizontally growing superimposed coral colonies. Seaward of shallow (4 to 5 metres) sandy or muddy depressions, pinnacles with rich coral growth and steep or overhanging sides rise to the same level as the reef flat.

Another fringing reef without a boat channel has been described from the Sarso Islands in the southern Red Sea (Scheer, 1971). These islands are emerged reefs, probably originating as ridge reefs, common in the Red Sea (see section 5 of this chapter). The fringing reef is very narrow, below a coastal corrosion notch in the emerged fossil reef, and the beach is restricted to a thin sheet of sand and silt on the lower part of which sea grasses grow. Alternations of coral heads and small hollows with a shallow sand fill extend as far as the outer edge, but there is no boat channel.

In rare instances, reefs growing offshore in coastal waters can become attached to the mainland by progradation of the coastline behind them. This has happened near Yule Point, North Queensland, where, under a humid tropical climate, rivers discharge large quantities of muddy sediment which is the cause of the progradation (Bird, 1971). In such conditions, the absence of boat channel occurs in a climatic environment just opposite to the above mentioned Red Sea reefs. The existence of such reefs is obviously threatened by excess of sedimentation.

1.2. Fringing reefs with an incipient boat channel

This subtype is represented 6.75 kilometres south of Aqaba harbour, i.e. 750 metres south of the reef shown on Fig. 68a. According to Mergner and Schuhmacher, the only difference (Fig. 68b) is the presence here of a very shallow depression, 1.50 metres below the general surface of the reef flat. This 'lagoon' does not contain much sediment, but the coral growth is different in that it consists of numerous scattered colonies. The difference could be due to local minor variations in sediment supply.

On the other side of the same gulf near Elat, Israel, an incipient boat channel is also found in the fringing reef (Mergner, 1971). The beach is fed by granitic sand and stones, mixed with calcareous sand and including beach rock slabs. It is bordered by a boat channel, not more than 2 metres deep, and between 10 and 40 metres wide, with a sandy and muddy bottom from which grow coral heads of various shapes and sizes. The reef flat is 15 to 30 metres wide and 0.3 to 0.8 metres deep, with a complicated pattern of deeper holes partly filled by coral detritus. Larger coral colonies occur on the outer edge with the usual horizontal growth. Detrital material appears again at the foot of the slope, but it is largely covered by living corals as the depth is still very small. Knolls exist in front. Such reefs on both sides of the innermost parts of the Gulf of Aqaba or Elat are obviously young, being built over the apron of apparently Pleistocene material which covers the crystalline basement.

Farther south in the Gulf of Elat along the Sinai Peninsula, fringing reefs display more complicated patterns and have had a longer history (Guilcher,

1979a). They include remnants of sharms or gullies cut during the Pleistocene regressions (Fig. 38), and are backed by sebkhas and mangrove swamps. Relics of former fringing reefs outcrop locally at a similar level.

Mahé Island in the Seychelles has many fine fringing reefs, with shallow boat channels containing rippled sands (Lewis, 1969). Behind these, meadows of marine grasses such as *Thalassia* and *Cymodocea* are exposed, a feature which is frequent in the Western Indian Ocean (e.g. in Madagascar and Kenya). The outer parts of these fringing reefs include ridges formed by dense growths of brown algae, which are replaced by branching corals on the outer sides of windward reefs.

Fringing reefs with discontinuous boat channels and fringing reefs without boat channels alternate along the southwestern coast of Sri Lanka at Hikkaduwa near Galle, where they have been illustrated by Mergner and Scheer (1974).

1.3. Fringing reefs with a well-developed boat channel

Lord Howe Island in the Tasman Sea between Australia and New Caledonia bears the southernmost reefs in the world, and provides a good example of a fringing reef with a distinct boat channel (Guilcher, 1973; Slater and Phipps, 1977; Veron and Done, 1979). As Fig. 69 shows, the fringing reef extends over 6 kilometres along the west coast of the island. It is divided into several segments by passages, which are shallow except for Erscott's Passage (15 to 19 metres). The outer slope does not descend immediately to great depths (Lord Howe lies on an insular shelf, 30 × 18 kilometres wide at a depth of 50 metres), but displays well-developed spurs and grooves. The boat channel, locally known as the lagoon, is shallower than one metre at low spring tide, but distinctly below the level of the reef which is fully exposed under these conditions. The reef includes many living colonies, especially *Acropora* spp., but also a number of soft corals (*Alcyonaria* spp.), and algae such as *Caulerpa* and *Padina* spp. In the north part the reef continues into the boat channel as a series of arms which have grown in alignments recurved by the refracted southwesterly oceanic swell, and consisting of dead coral blocks of medium and small size. The boat channel ends landward in a beach of calcareous sand, with coral and shell debris, backed by recent dunes and consolidated calcareous aeolianites. The climate is subtropical with Mediterranean affinities in the monthly distribution of rainfall. The aeolianites behind the beach have been cut back to form a shore platform in front of an overhanging cliff, with mushroom-shaped stacks rising from the platform (Fig. 70). In the boat channel the sediment is of variable thickness, and fills buried valleys that have been discovered beneath its floor. It seems that the present (Holocene) fringing reef rests upon a Pleistocene one, which has been eroded during the low sea-level phase(s). The aeolianites probably formed during the last marine regression as they continue below present sea-level.

In the Caribbean, a much narrower fringing reef with a boat channel may be mentioned at Jacmel on the south coast of Haiti (Fig. 71). This reef is

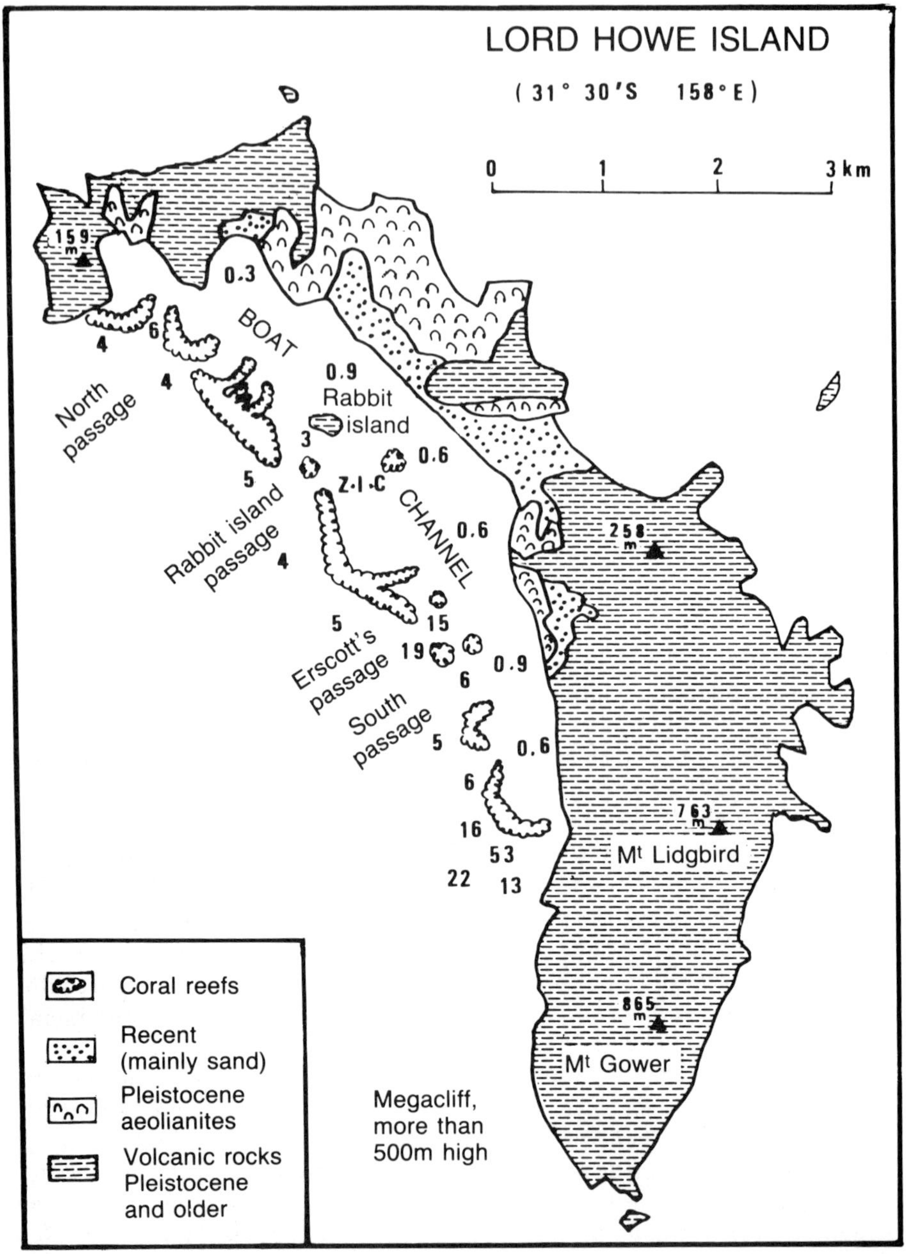

Fig. 69 Lord Howe Island, Tasman Sea, and its fringing reef. After Guilcher (1973) (R V *Mendele'ev* expedition, 1971). Depths in metres below low spring tide from marine charts

Fig. 70 Inner edge of Lord Howe fringing reef, Tasman Sea, cut in Pleistocene aeolianites. Mt Gower megacliff (more than 500 metres high, see Fig. 69) cut in volcanic rocks in the distance (Photo A. Guilcher)

Fig. 71 Narrow fringing reef at Jacmel, south coast of Haiti, Hispaniola, Caribbean, with shallow boat channel bearing Phanerogams on sandy bottom (Photo A. Guilcher)

backed by a series of elevated coral terraces (Haiti belonging, as said in Chapter three, section 7, to the coasts girdled by sets of raised reefs). Seaward from the coastline is an inner platform with big blocks that have fallen from the overhanging cliff cut into the previous reef; a narrow but well-shaped boat channel with phanerogams growing on a sandy bottom; and an outer reef

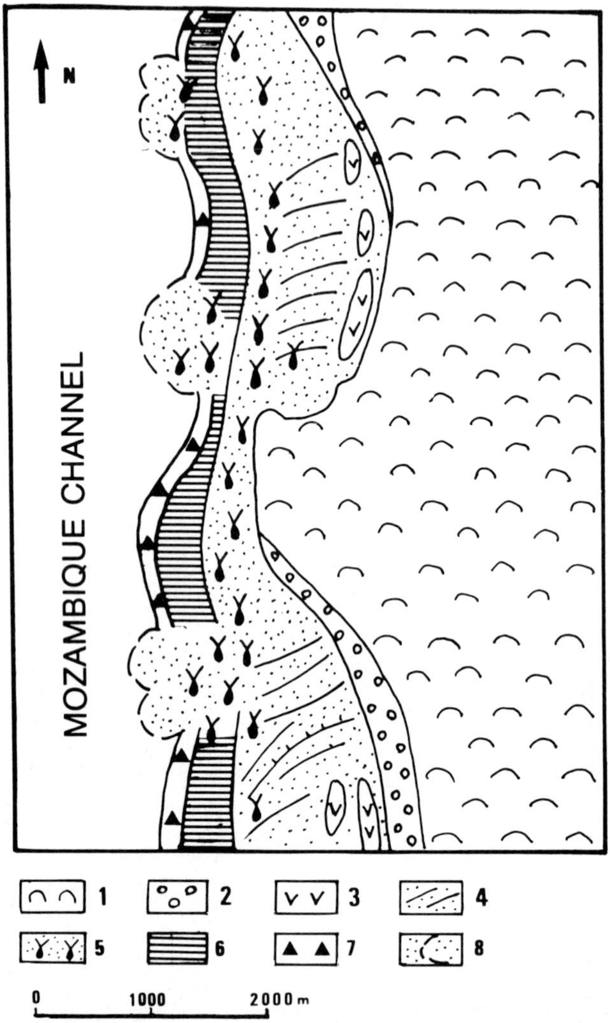

Fig. 72 Fringing reef on the southwestern coast of Madagascar (lat. 24°S). Simplified from Battistini, 1964. 1: Pleistocene and Holocene dunes; 2: beach; 3: Phanerogam meadows in inner boat channel; 4: boat channel with striations in sand; 5: scattered coral colonies; 6: reef flat; 7: detrital outer ridge with big blocks; 8: submarine sandy delta

bearing many sargassoes. Scattered path reefs grow in front, as in the examples quoted from the Red Sea.

The fringing reef on the southwest coast of Madagascar (Fig. 72), described by Battistini (1964) and Pichon (1972), has grown in a much higher wave energy environment since this coast is exposed to a southwesterly swell generated in the Southern Ocean. It is advisable here to proceed from sea to land in the description. On its outer side the reef is capped by a detrital outer ridge including big blocks (the so-called negro heads). This ridge is interrupted by gaps filled by submarine sandy deltas generated by strong ebb currents, the tidal range being 3.3 metres at spring tides. The detrital accumulations in incisions in the reef are thus in a reverse direction to those in Lord Howe fringing reef (see above). Inside the outer ridge the reef flat is exposed at low spring tides; it is backed by a boat channel extending to a depth of 1 to 3 metres below the reef flat. Coral growth is very active on the sandy bottom in the outer part of this channel, while in the inner part the sand is in parallel rows, as shown on Fig. 72. Some phanerogam meadows exist near the beach. In such a fringing reef, the existence of a boat channel does not seem to be mainly related to a sediment supply from the mainland, but rather to the damming effect of the outer ridge, which is itself a consequence

Fig. 73 Fringing reef at Nosy Valiha, an islet off the northwestern coast of Madagascar. Low spring tide. From sea to land: outer edge with spurs and grooves; outer ridge; striated reef flat; shallow boat channel with Phanerogam meadows associated with crescentic ponds (as those shown on Fig. 23); mangroves; sandy beach (Photo A. Guilcher)

of the surf; the sand rows in the boat channel could be related to surf beats, or to tidal currents.

Figure 73 gives an example of another complex fringing reef zonation, also in Madagascar, but in a less energetic environment. Boat channel reefs exist at La Réunion, Indian Ocean (Montaggioni, 1974).

1.4. Fringing reefs with multiple landlocked mini-lagoons

These reefs of unusual shape have been identified on the northwest coast of Madagascar, in Ramanetaka Bay at 14°20′S (Guilcher *et al.*, 1958; Guilcher,

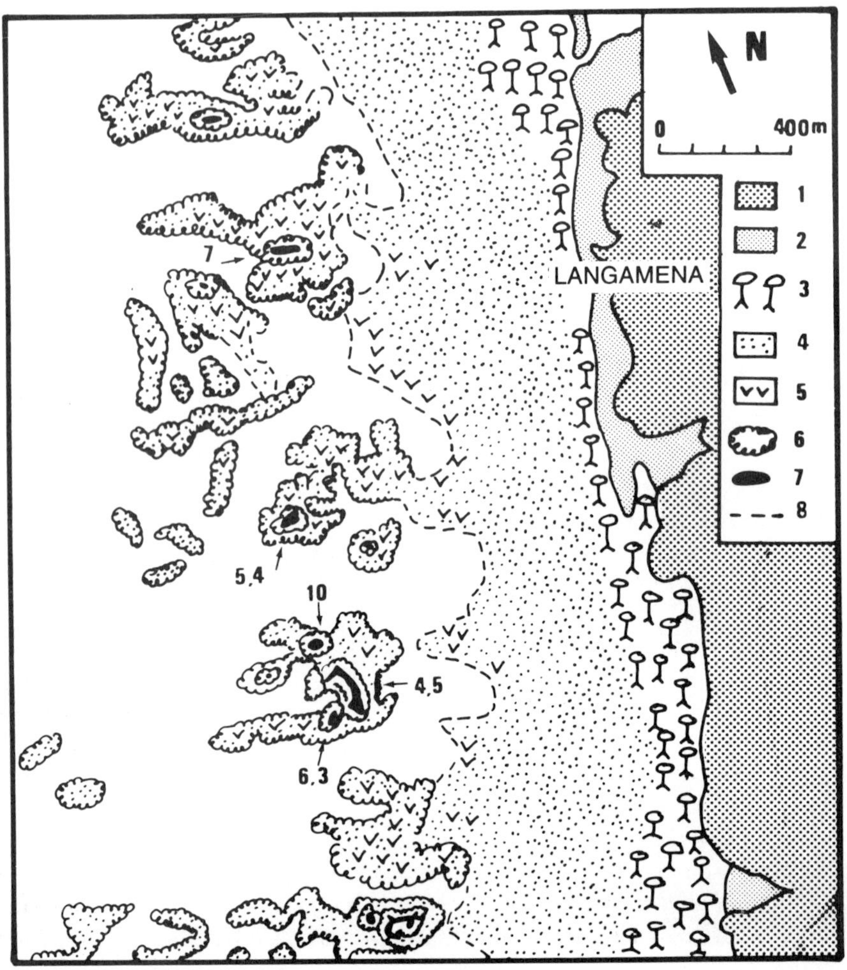

Fig. 74 Fringing reefs with multiple landlocked mini-lagoons off Langamena village in Ramanetaka Bay, Madagascar (lat. 14°20′S). After Guilcher, Berthois, Battistini and Fourmanoir (1958). 1: hills; 2: coastal plain; 3: mangrove; 4: silt, sand, gravel, drying at low spring tide; 5: Phanerogams (Cymodoceae) meadows on reef flat; 6: peripheral rim around reefs; 7: small landlocked lagoons. Figures with arrows: lagoon depths in metres at low spring tide; 8: low spring tide line

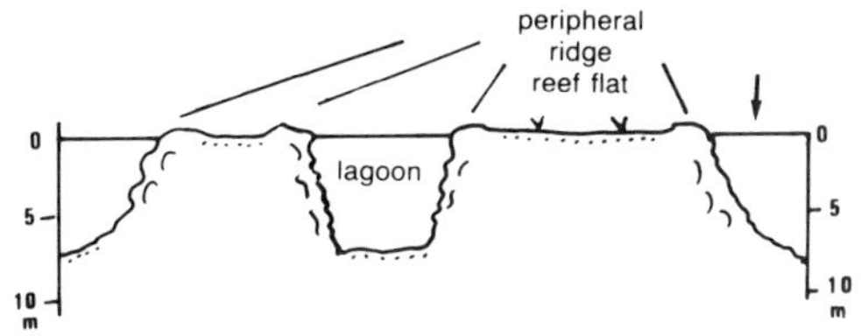

Fig. 75 Section across a reef in Ramanetaka Bay. After Guilcher *et al.* (1958)

1959). The swell travels over a long fetch, but is very moderate, in contrast with the area shown on Fig. 72. The tidal range probably exceeds 4 metres at spring tides, so that tidal currents are strong, and there is a good circulation of water in the bay. In front of Langamena village, the sequence from land to sea is as follows (Fig. 74).

Along the coastal plain a nearshore mangrove zone grows over 100–200 metres in width on a sandy and gravelly bottom (this is a common feature in the inner reaches of fringing reefs). Farther offshore, over 400–500 metres, the bottom is still predominantly sandy, with 10–15 per cent of silt. *Cymodocea* meadows appear in places low spring tide level.

The reefs themselves consist of long arms of coral, whose growth has certainly been rapid, favoured by the good circulation of water due to tidal currents. These digitate corals meet in many places, so that they enclose parts of the sea floor, which in this area is 8 to 12 metres deep. The result is that inside the coral ridges are small enclosed lagoons, 5 to 10 metres deep at low spring tide, completely sealed but connected with the open sea by their surface except at low spring tides, as shown on Fig. 75. The reefs are surrounded by rims of *Porites* and *Astraea*, their inner parts (reef flat on Fig. 75) bearing sand, gravel, phanerogam meadows, and small coral colonies. The sediment in the mini-lagoons is finer, including 50 per cent of silt and clay, because of their sheltered situation. Still farther offshore, the reef growth continues in the shape of coral knolls or patch reefs. Outside the area shown on Fig. 74 these coral knolls are extremely numerous. In some places several kilometres offshore the knolls have also enclosed mini-lagoons as those on Fig. 75.

Summarizing the facts, the Ramanetaka case seems to owe its features to a very unusual environment: water stirred by strong tidal currents; moderate surf, so that waves do not damage the reefs; moderate depth (10–18 metres) of the bay, so that reefs soon grow to the surface; and absence of pollution, since the mainland is very sparsely populated: Langamena is but a tiny village.

1.5. Reefs intermediate between fringing and barrier

These reefs may be exemplified by some of the raised reefs described by Chappell (1974) on the Huon Peninsula in New Guinea (Fig. 76). Some of

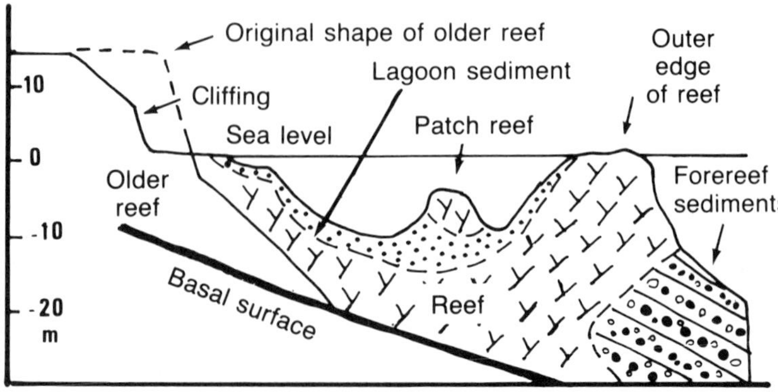

Fig. 76 Fossil raised reef of the Huon Peninsula, Papua New Guinea, intermediate between fringing and barrier reef. No scale for lengths. Simplified from Chappell (1974)

these emerged reefs had their outer part separated from the land (i.e. from the earlier reef) in the course of the general uplift, by a stretch of water deeper than the usual boat channel (10 metres in Fig. 76). In such a situation the term lagoon seems more appropriate than the expression boat channel. The former 'lagoon' zone is easily recognized from its sediments, different in texture from those of the built-up reef. Patch reefs can exist in such a lagoon, with a greater height than the coral heads scattered in boat channels.

This is a fossil reef, but such an intermediate subtype is also represented in living reefs. Along the Kenya coast (Bird and Guilcher, 1982), the reef belongs really to the fringing type is the surroundings of Mombasa. Farther north, however, near Malindi, it passes laterally into a more complex reef, 3 kilometres wide, including a landward zone of phanerogam meadows and coral colonies on sandy bottom, two or more areas several metres deep with interspersed shallows, and finally a very shallow outer reef. Near Port Sudan in the Red Sea (Schroeder and Nasr, 1983), the fringing reef includes discontinuous depressions, 20 to 25 metres deep, which could have been produced by karstification during a low sea level phase: in which case the fringing reef rests at depth on a dissected Pleistocene reef.

In the Caroline Islands and the Marianas, several examples of intermediate situations have been described by Tayama (1952) at Kusiae Island (p. 41 and 233), at Tinian Island (p. 89–90), at Merizo, Guam Island (p. 91). He called all these structures almost barrier reefs. At Kusaie and Merizo there are elongated sand cays on the reef.

Such transitional forms indicate that the distinction between fringing and barrier reefs is by no means absolute, although in the majority of sites discrimination is easy.

1.6. Asymmetrical fringing reefs

On islands where the direction of wave attack is well defined, with much stronger wave action on the exposed shore than on the sheltered side, it happens, when a small rocky island stands offshore with moderate depths around it, that a fringing reef grows on the exposed side only, whereas on the sheltered side a sandy tail is built. Fig. 77 shows an example of this at Nosy Iranja off the northwest coast of Madagascar (Guilcher, 1956). The swell comes from the northwest as in Ramanetaka Bay (section 1.4, above), where this subtype is also represented on four other small islands (Guilcher, 1956; Guilcher *et al.*, 1958). The dynamics of such frameworks have some resemblance to the glacial forms known as crag and tails in Scotland. In the case of coral islands such as Nosy Iranja, the absence of a fringing reef on the leeward side seems due to the arrival of an excess of sediment supplied by erosion on the windward side. An emerged sand cay with beach rock indicates that Nosy Iranja belongs to a somewhat mature stage. Comparison should be made with the sand cays associated with isolated reef patches in large lagoons as behind

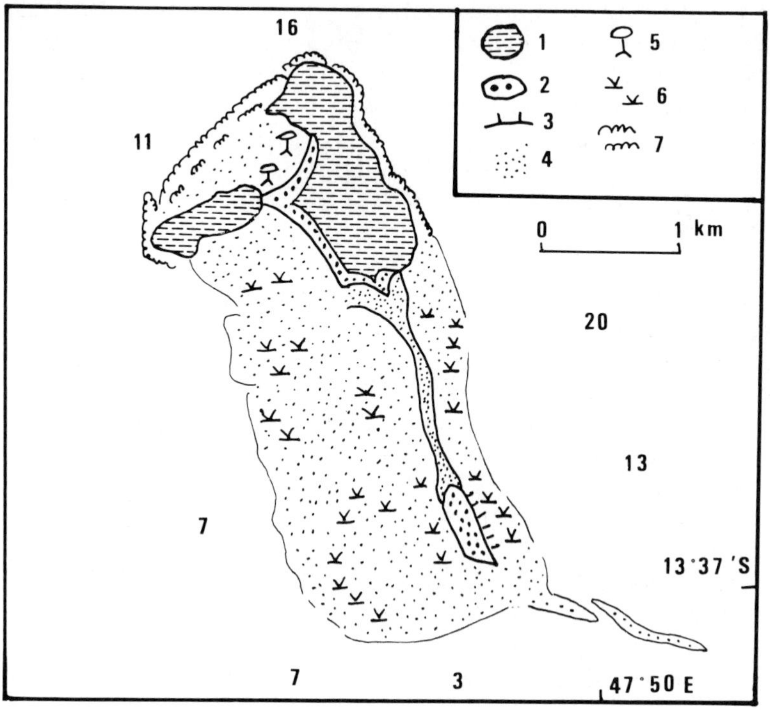

Fig. 77 Nosy Iranja, off the northwestern coast of Madagascar, an island with an asymmetrical fringing reef. After Guilcher (1956). 1: rocky core of the island; 2: beach and sand cay, both emerged; 3: beach rock; 4: intertidal sand. 5: mangrove; 6: phanerogam meadows; 7: coral reef. Depths in metres below lowest water mark

Fig. 78 Falcon Island and Brisk Island, Great Barrier Reef, looking south-east. Fringing reefs on windward side (left and background); sand tails on leeward side (right and foreground). The islands themselves are not made of coral reefs (Photo E. C. F. Bird)

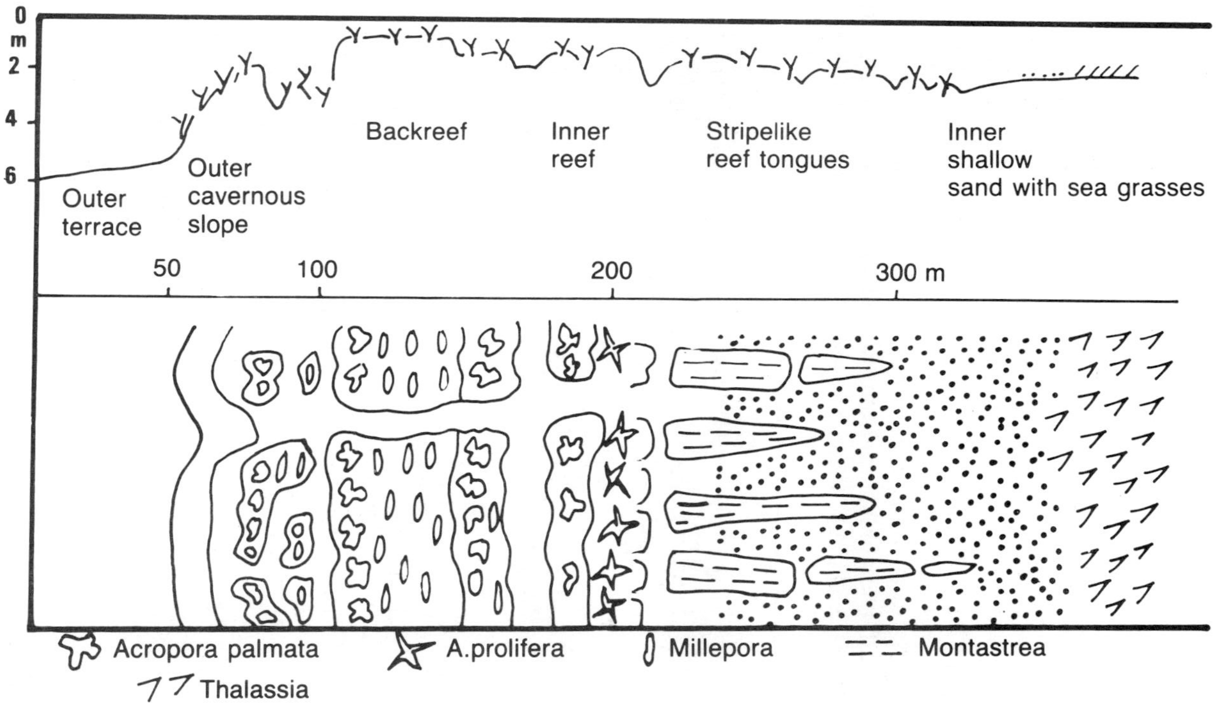

Fig. 79 Section and plan of Cayo Macao bank reef, Gulf of Batabano, Cuba, showing dominant species or genera. After Kühlmann (1971)

the Great Barrier Reef and in New Caledonia, but without emerged rocky islet to support the structure (section 3.6 of this chapter; two other examples are given on Fig. 78).

2. BANK REEFS

Under this name, a number of reefs may be grouped. The smallest are often called *patch reefs*. They rise from various depths, and lie at various distances from the land. Their shapes can be quite different. Off the coasts of Cuba, numerous small bank reefs have been carefully described by Kühlmann (1970, 1971, 1974). At Cayo Macao Reef in the Gulf of Batabano for example (Fig. 79) the usual zonation from exposed to sheltered side appears clearly. The depth from which the reefs grow is very small. The other examples described by Kühlmann from various coasts of Cuba are not essentially different, although some are more like fringing reefs (e.g. at

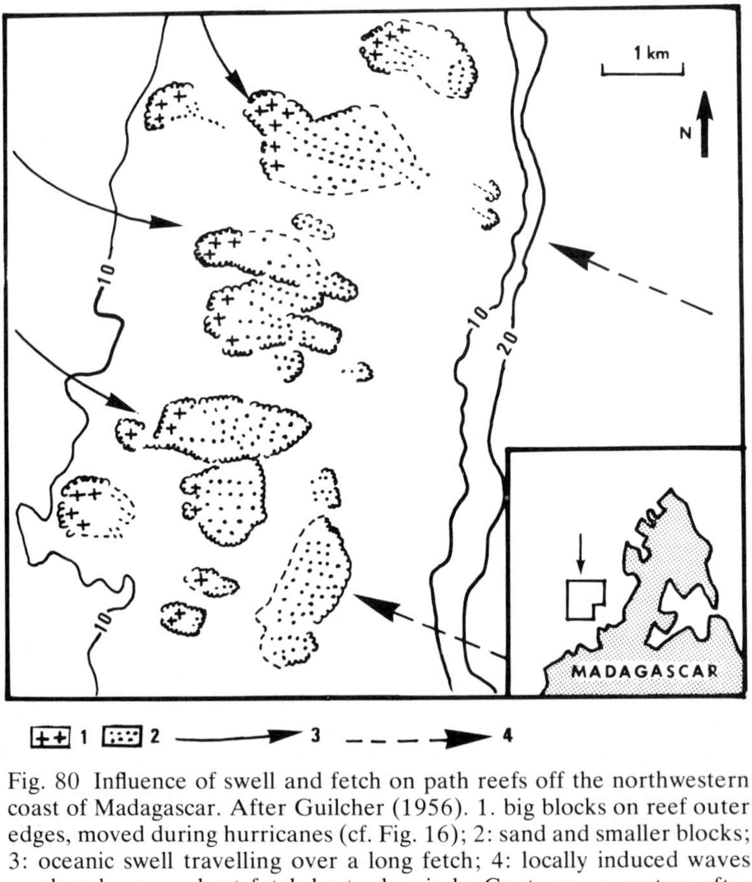

Fig. 80 Influence of swell and fetch on path reefs off the northwestern coast of Madagascar. After Guilcher (1956). 1. big blocks on reef outer edges, moved during hurricanes (cf. Fig. 16); 2: sand and smaller blocks; 3: oceanic swell travelling over a long fetch; 4: locally induced waves produced over a short fetch by trade winds. Contours on metres after French chart

Rincon de Guanabo, *op. cit.*, 1970, p. 735, where the outer slope descends to depths exceeding 50 metres). Near La Vera Cruz in Mexico, Kühlmann (1975) described other bank reefs, generally with steeper slopes but also rising from shallow depths, and with contrasting leeward and windward sides.

An interesting case of patch reefs was reported by Guilcher (1956) off the northwest coast of Madagascar (Figs. 80 and 81). These reefs, rising from shallow depths, are exposed to an oceanic swell arriving from the northwest, generally moderate but occasionally strong during tropical cyclones. On the eastern side, the reefs are washed by a gentle swell generated by the trade winds, very steady but travelling over a very short fetch, as the mainland is close behind the reefs. Consequently, on their western edges the reefs bear large blocks displaced by waves during cyclones, whereas on their eastern sides much finer sediments have settled under the influence of the easterly waves. In other words, these patch reefs display a double zonation, as they have two unequally exposed sides instead of one sheltered side.

Other bank reefs are those described by Teichert and Fairbridge (1948) on the Sahul Shelf off the northwest coast of Australia. They are small, up to a few kilometres long, often almost circular, and rising from depths ranging from 50 to 115 metres. They often include sand cays on the sheltered side, and thus resemble more or less the 'low isles' in lagoons behind barrier reefs (see section 3.6). They give place to atolls on the outer edge of the shelf (Berry and Marsh, 1985). According to Van Andel and Veevers (1967), the

Fig. 81 Patch reefs built in opposite directions, part of Fig. 80, as seen from the east. Low spring tide (Photo A. Guilcher)

Sahul Shelf has been subject to slow orogenic response to the disturbances occurring in the bordering Indonesian geosyncline. In Late Cenozoic it was subsiding, but apparently became stable during the Quaternary. The formation of its bank reefs has not yet been investigated in detail.

Bank reefs with sand cays occur on the open shelf off southwest Madagascar, as at Nosy Andriangory (Salomon, 1980) (Fig. 102e). They are exposed to heavy swell generated in the Southern Ocean, and bear huge blocks on their outer ridge, but their exposed situation does not prevent the leeward cay to be fully mature.

Beside these bank reefs, built in shallow areas, others rise from great depths. The best examples in the world are perhaps those scattered in the Coral Sea, northeast of the Great Barrier Reef (Fig. 82). There are more than

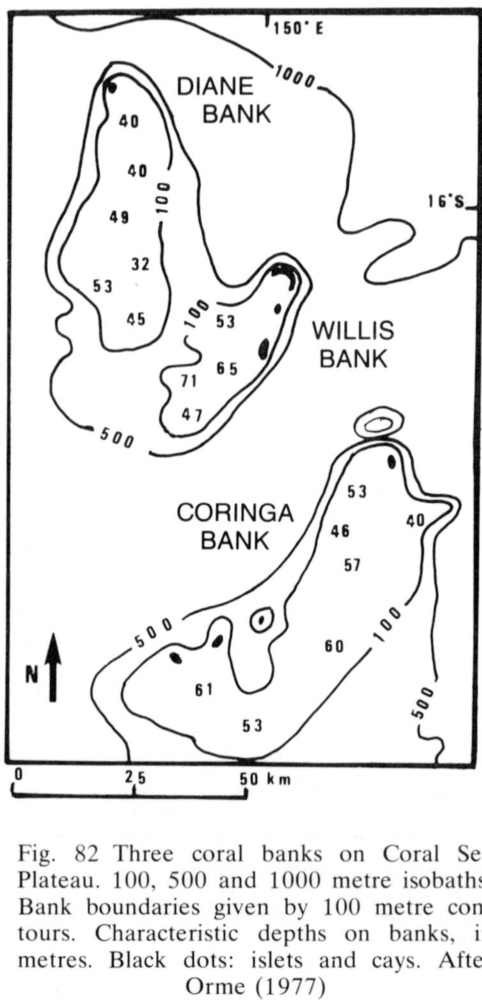

Fig. 82 Three coral banks on Coral Sea Plateau. 100, 500 and 1000 metre isobaths. Bank boundaries given by 100 metre contours. Characteristic depths on banks, in metres. Black dots: islets and cays. After Orme (1977)

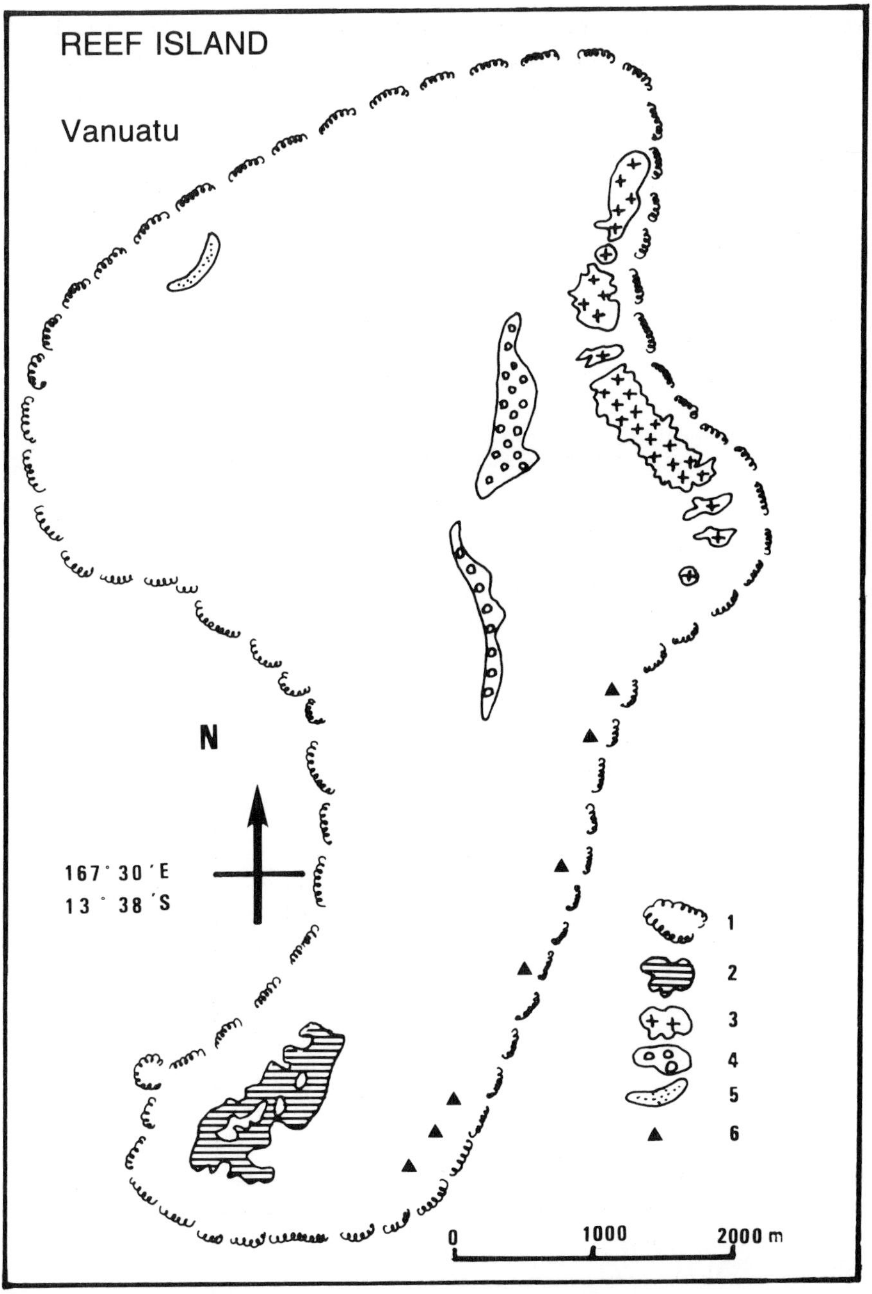

Fig. 83 A bank reef: Reef Island, Vanuatu, Melanesia. After Guilcher (1974). 1: bank reef at sea-level: basement probably Pleistocene with thin sheet of loose sediment; 2: karstic hole with pool, carved into 1; 3: islets (rampart) of Middle Holocene age (see Fig. 84); 4: sandy windward islets; 5: leeward sand cay; 6: big detrital blocks thrown by hurricanes onto windward edge

twenty of these coral banks (Orme, 1977), of different size. The largest one, Tregrosse Bank, is 95 **g** 50 kilometres, being thus larger than all atolls except perhaps Kwajalein Atoll in the Marshall Islands (120 **g**32 kilometres). Several others exceed 30 kilometres. They rise steeply from a wide plateau, 600–2000 metres deep, tilted from southeast to northwest, which forms an intermediate step between the Great Barrier Reef in the southwest and the Coral Sea Basin in the northeast where there are depths of more than 4500 metres. The relatively flat surface of these banks is at depths of some tens of metres (usually 50–60 metres), the viable reefs being shallower and mainly found near the bank margins. A few reefs bear cays and islets, both vegetated and unvegetated, so that some banks such as Tregrosse, Coringa, and Willis more or less resemble huge drowned atolls. The plateau on which they stand is thought to be a result of a Tertiary differential subsidence, which separated it from the Queensland continental shelf upon which the Great Barrier Reef is built. The continental nature of the crust here has been verified by seismic reflection. The formation of bank reefs on this plateau needs still further research.

The large Chagos Bank in the Indian Ocean, even larger than those of the Coral Sea, has been mentioned in Chapter three, section 5. Several smaller bank reefs in the same area, as Coetivy, Cargados Carajos, and Tromelin, still await investigation.

Another small (8.5 × 5.2 kilometres) bank reef, called Reef Island, and located in the Banks Islands, northern Vanuatu (Fig. 83), is an example of the

Fig. 84 Reef Island, Vanuatu. Rampart of Middle Holocene age on windward side (symbol 3 on Fig. 83), consolidated and cliffed, with visor (Photo A. Guilcher)

diversification of landforms which can occur in this type of structures (Guilcher, 1974). It consists of a coral platform at sea-level, probably Pleistocene, and karstified during the Wisconsinian regression, hence the hollow 7 metres deep in the south. It bears three types of low islets: from east to west, a windward series of slightly emerged Middle Holocene conglomerate (Fig. 84); a central sandy series, also facing the trade winds; and a leeward sand cay.

Many small bank or table reefs, rising to the sea surface from great depths, occur in the Caroline and Mariana Islands, as Losiep, Merir, and Gaferut (Tayama, 1952, p. 114–120 and 261–265). In the same area, Tatsumi Bank (*ibid*., p. 123 and 264) is a table reef drowned under 7 metres of water in its shallowest part. At Saipan and Guam, tops of hills or mountains have been interpreted by Tayama (p. 264) as raised fossil table reefs. In the Caribbean off the shelf of Nicaragua, San Andres Island is a raised bank reef, uplifted in Post-Miocene times, which is thought to have a deep-seated volcanic basement (Milliman and Supko, 1968).

3. BARRIER REEFS AND ASSOCIATED FEATURES

3.1. Various barrier frameworks

Barrier reefs can be continuous, with only rare passages, or fragmented into small units. Examples of continuous linear reefs occur at Tulear in Madagascar (Pichon, 1974; Weydert, 1974, 1979) and in the northern section of the Mayotte barrier (Fig. 85) for 18 kilometres without any break, with fine differences, however, between the oceanward and the lagoonward edges. The western and southern barriers of New Caledonia, and the Cook Reef at the northern end of the same island, provide many examples of rectilinear reefs (Fig. 86), as does the barrier encircling Raiatea and Tahaa Islands in the Society group, where almost all the existing passages are bordered on either side by twin sand cays (see photographs in Guilcher, 1985c, p. 1026, and charts in Davis, 1928, p. 295 and 297): this feature exists in other coral seas, and has been explained by Steers (1929) as the result of refraction of the swell on either side of the passage, thus supplying sediment. The Bora Bora barrier provides another example (Fig. 87). The long barrier reef bordering the southern side of Tagula Island, at the east of New Guinea, belongs to the same subtype, with only three passages within a distance of more than 120 kilometres. Linear reefs form barriers in the Fiji Islands, especially in the Lau Group (Exploring Isles, Nggamea near Taneuni, and others); in the Exploring Isles, the American Passage across the barrier is of quite exceptional depth, exceeding 100 metres.

A different pattern occurs where barrier reefs are divided into many segments, each comparatively small and recurved inwards at both ends, by the same process of wave refraction as in the twin sand cays just mentioned. This subtype is well illustrated in the Great Barrier Reef off Lizard Island

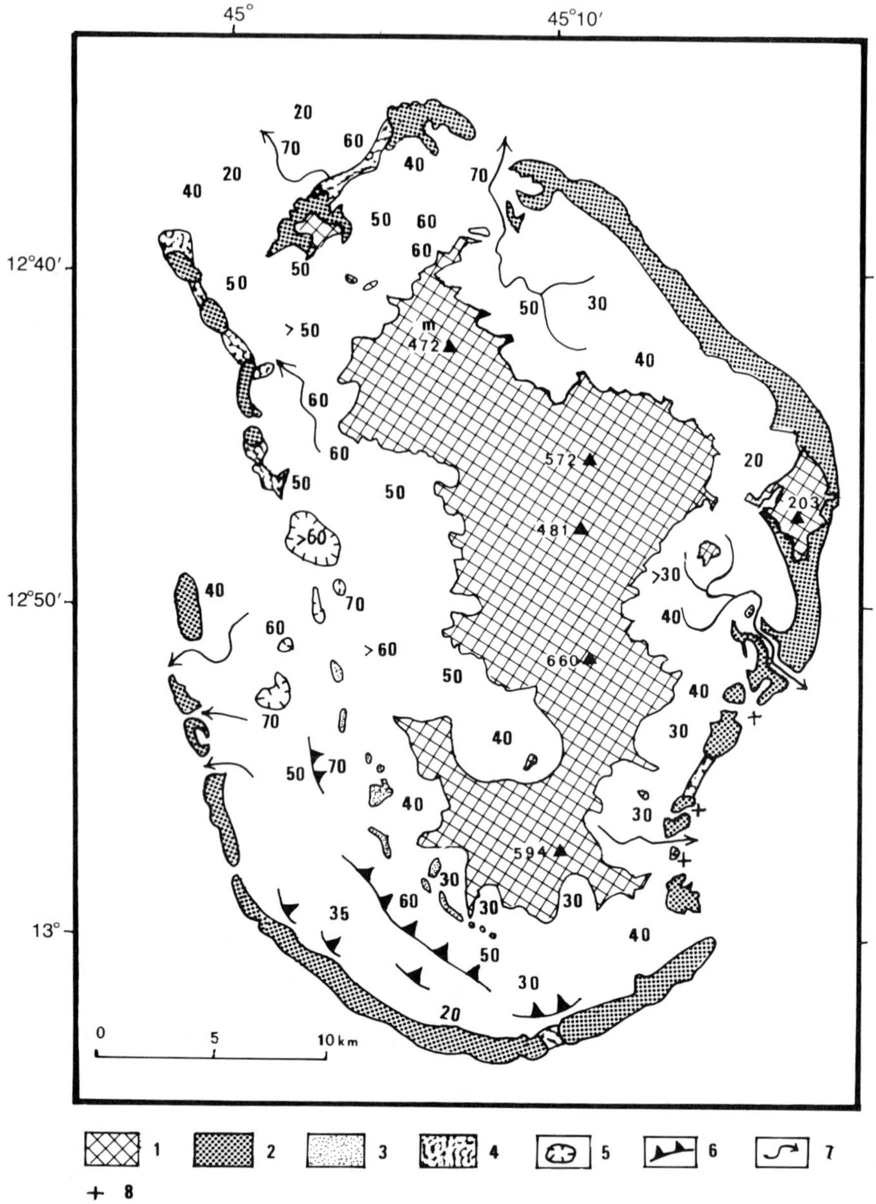

Fig. 85 Mayotte barrier reef and lagoon, western Indian Ocean. After Guilcher *et al.* (1965) with a few additions from Coudray *et al.* (1985). 1: volcanic mountains with heights in metres (the eastern one, 203 metres high, has been recently intruded through the barrier); 2: outer, main barrier; 3: inner barrier (in southwest only); 4: parts of barrier drowned under several metres of water; 5: probable karstic depressions (sink holes) in the lagoon; 6: faults with downthrow; 7: drowned Pleistocene valleys; 8: infilled passage on windward side. Approximate depths in metres below low spring tides

Fig. 86 Southwestern part of New Caledonia barrier reef, Melanesia. Rectilinear with only rare passage (Photo A. Guilcher)

Fig. 87 Bora Bora Island and barrier reef, Society Islands, as seen from the west. Teavanui (i.e. the Great Passage) located on the leeward side (compare with locations of avas or passages in the western Tuamotus, Fig. 34), with sand cays on either side as result of swell refraction (Photo A. Guilcher)

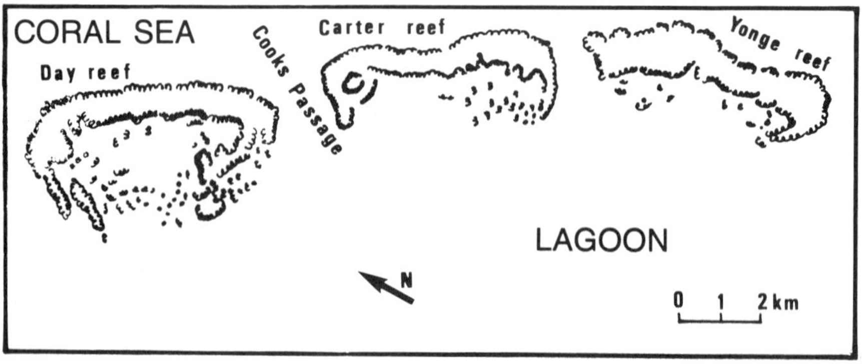

Fig. 88. Arcuate reefs forming elements of the outer edge of the Great Barrier Reef at latitude 140°30′S, where the lagoon is about 65 kilometres wide

(14°30′S) and off Cooktown (15°10′–15°30′). Fig. 88 shows this subtype, which has been carefully analysed and pictured by Fairbridge (1950, p. 338–341) and called by him and others ribbon reef, although this name would perhaps better fit the first subtype. The inner side of these reefs forms a sheltered area, in which a large number of tiny coral knolls grow. At Day Reef (Fig. 88), the recurves partially enclose a small lagoon. Other farolike features of this shape, more advanced in development, have been described from several parts of the Mayotte barrier (Guilcher, 1965, 1971; Guilcher *et al.*, 1965; in one site, the small lagoon is completely enclosed (Figs. 89 and 90).

Fig. 89 Mayotte barrier reef, western side: a faro-like reef which secundary lagoon enclosed as a result of wave refraction at passages on both sides (Photo R. Battistini)

The structure of barrier reefs can also differ. In the barrier reef which girdles Martinique Island in the Carribean, more than 50 kilometres long on the eastern side, exposed to strong swell generated by easterly winds (Adey *et al.*, 1977; Battistini, 1978), corals are replaced on the crest by a carbonate pavement covered with dense stands of fleshy algae, especially *Sargassum*, not only on the inner slope, but sometimes also on the upper part of the outer slope. Broad phanerogam meadows grow in the inner part of the lagoon, where fringing reefs are also common.

Several barrier reefs in the Society Islands have an inner slope invaded by sediments washed over the reef flat by the surf and entering the lagoon as protruding fans or aprons. The largest of these occurs at Maupiti (Fig. 91); others are visible on the north side of Tahaa barrier reef, and there are several on Bora Bora. In the north of Bora Bora there is a combination of recurved segments of the barrier, convex outwards, which have emerged as wooded islands, and shallow fans, convex inwards and encroaching into the lagoon (Fig. 92). The fans are fed with sediment washed through shallow channels called *hoas* during periods of heavy swell. These hoas are described in more detail later in the Tuamotu atolls (this chapter, section 4.6).

Palau Island in the Western Pacific has fringing reefs in the northeast and east and barrier reefs elsewhere (Tayama, 1952, p. 237 and Fig. 8 in Vol. 2). The same change occurs around Tagula Island near New Guinea, and Rambi Island in the Fiji Islands. Similarly, Guam Island in the Marianas has fringing

Fig. 90 Inner side of Mayotte barrier reef. Shallow secundary lagoons enclosed and invaded by coral growth. The main lagoon is behind (Photo A. Guilcher)

Fig. 91 Maupiti barrier reef, Society Islands (see Fig. 54). Sediment has been washed into the lagoon through two hoas (shallow passages) to form a delta (Photo A. Guilcher)

Fig. 92 Bora Bora Barrier reef, Society Islands, eastern side. The barrier is divided into crescentic segments separated by hoas (shallow passages) through which sediments are washed into the lagoon (Photo A. Guilcher)

reefs along its coasts, except for the southernmost tip, where a barrier reef is backed by a lagoon, this structure being some kilometres wide (Tracey *et al.*, 1964).

The interpretation by Purdy (1974) of the barrier reef off Belize, Caribbean, as a result of karstification during lower sea level, has been noted in Chapter three, section 5.

3.2 Sediments in lagoons behind barrier reefs

The nature of sediments in lagoons behind barrier reefs indicates the ratio of supply from the central island or mainland and the supply derived from corals and associated marine organisms. This has been investigated in several lagoons in the Indian and Pacific Oceans (Mayotte, New Caledonia, Bora Bora, Tahiti) and in the Caribbean (Guadeloupe), by means of the proportion of $CaCO_3$ in bottom samples, an adequate criterion where the central island is siliceous (Guilcher *et al.*, 1965a,b,c, 1966, 1969; Guilcher and Marec, 1978; Dugas and Debenay, 1980, 1981, 1982; Debenay, 1987; for general results, Guilcher, 1985a). The conclusion was that, in the humid tropical environments where all these lagoons lie, the percentage of terrigenous and skeletal particles depends mainly on the relationship between the areas of the central island and the bordering lagoon. There were significant contrasts between Bora Bora and Tahiti, on the one hand, and between the two lagoons surveyed around New Caledonia on the other. Three examples are shown on Figs. 93, 94 and 96. The general distribution in Tuo lagoon, New Caledonia (Fig. 96) is similar to that in the lagoon behind the Great Barrier Reef at latitude 14°30′–15°S (Orme *et al.*, 1978a), where terrigenous sedimentation from the Australian continent and high islands dominate the inner part, while the influence of *Halimeda* gravels and sands is dominant in the outer part.

However, such distributions have been recently modified by man, in places where the supply of sediment from the land has been increased by the cultivation of steep slopes, leading to an increase of lagoon silting by rivers, and a decrease of the $CaCO_3$ proportion in the lagoon. This has happened at Mayotte (Guilcher, 1985a) and is part of the problem of reef pollution which will be discussed in Chapter five (Fig. 150).

3.3. Drowned barriers

At Madagascar, Guilcher (1954) has described a drowned barrier which extends for hundreds of kilometres along the northwestern coast. At the northernmost tip of Madagascar it continues as an emerged Pleistocene reef. In its submerged part it bears only a few coral colonies, although the depth does not exceed 4 to 10 metres, and is thus quite suitable for coral growth. The 'lagoon' between the barrier and the mainland is much deeper, generally between 20 and 40 metres. The barrier is breached by deep passages off the main rivers, as in the case of true barrier reefs (maps in Guilcher, 1956, p. 111 and 113). The problems related to this drowned barrier (if it is really one, and

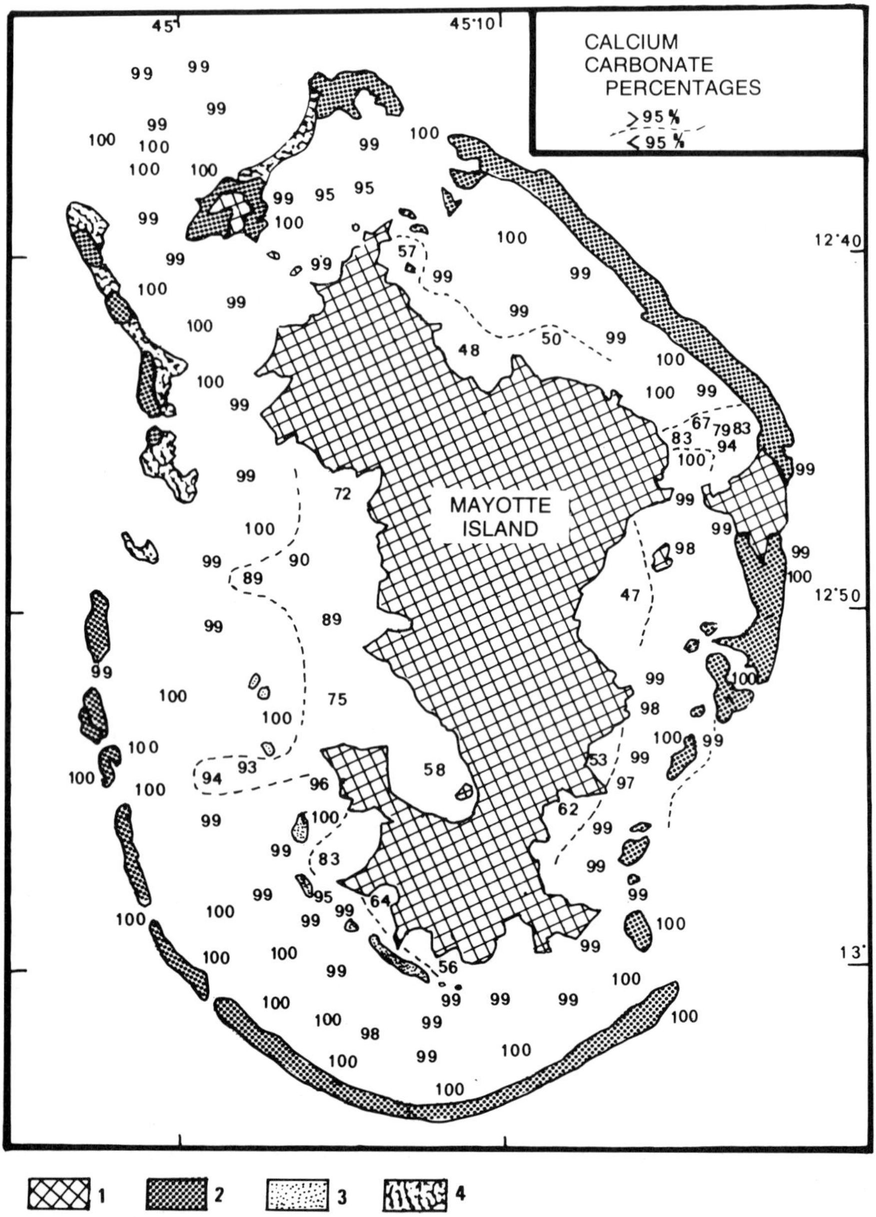

Fig. 93 Calcium carbonate percentages in Mayotte lagoon, Indian Ocean, in 1959. After Berthois in Guilcher *et al.* (1965). 1: volcanic island; 2: outer barrier; 3: inner barrier; 4: parts of barrier drowned under several metres of water

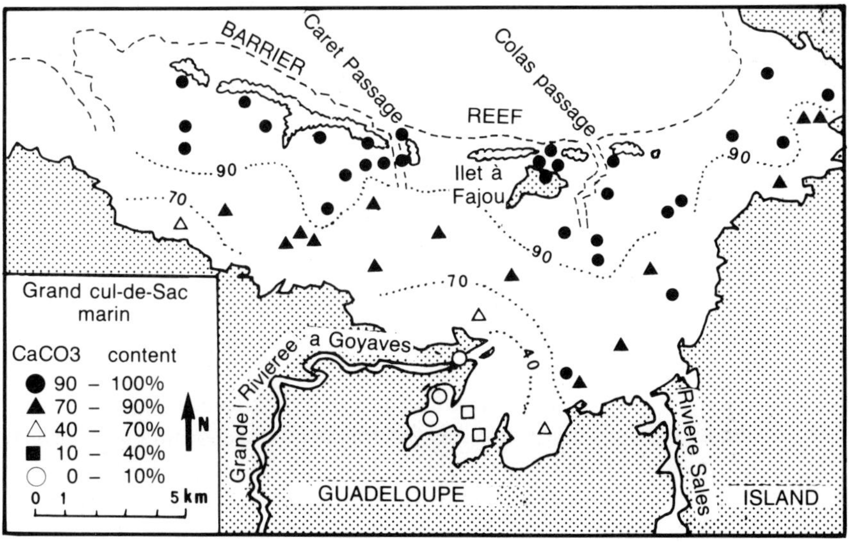

Fig. 94 $CaCO_3$ content of sediments in Grand Cul de Sac Marin, a lagoon behind the barrier reef of Guadaloupe Island, Carribean, showing the localized influence of Grand Rivière à Goyaves on sedimentation. Rivière Salée is not a river, but a marine strait. After Guilcher and Marec (1978)

not a structural feature) remain open so far. Also in the Indian Ocean, a partly drowned barrier reef lies in the Gulf of Marmar between India and Sri Lanka (Stoddart, 1973; map in Mergner and Scheer, 1974).

Another site where a drowned barrier reef exists as such is Tahiti (Guilcher, 1985c, Fig. p. 1024), where the barrier is submerged in three sections along the northeastern, eastern and southeastern coasts of the island, whereas it rises to sea-level in its other parts. Such features are understandable as adjustments of an island that has recently formed over a hot spot (Chapter three, section 6).

Another submerged barrier reef, mentioned by Davis (1928, p. 369) and shown on marine charts, extends along the southern coast of easternmost New Guinea for at least 160 kilometres. It is under 7 to 15 metres of water and is separated from the land by a lagoon 45 to 70 metres deep. It is a westward continuation of the ribbonlike barrier reef off Tagula Island, mentioned in section 3.1 of this chapter, which is not submerged. Two or three atolls extend in between. This area and the nearby Louisiade archipelago are places where further reef studies are needed and will certainly be most fascinating.

Still other submerged barriers are known in the Fiji Islands (e.g. in Davis, 1928, p. 315), an archipelago where they are adjacent to features described sometimes as raised barriers, but thought by Ladd and Hoffmeister (1945) and Purdy (1974) to be subaerial karstic features (see Chapter three, section 5). The general tectonic context makes these features quite understandable.

3.4. Almost-atolls

The finest almost-atoll in the world is Truk in the East Caroline Islands (Shepard, 1970; Tayama, 1952, p. 241–243 and Fig. 10 in vol. 2); Davis, 1928, p. 377–378). The reader should refer to the handsome chart published by Shepard. It is more or less circular or pentagonal, with a diameter of approximately 50 kilometres, and a circumference of 200 kilometres. About 20 high volcanic islands or islets rise above the lagoon surface, mostly on a central ridge running WSW–ENE. The largest of these islands, Tol, Moen and Fafan, are 7 to 9 kilometres long; the highest is 480 metres high. All have fringing reefs, some have mangroves on the inner reaches of their shores. The outer barrier reef is cut by about 50 passages; it is more continuous in the east than elsewhere; it is absent or drowned for only 5 kilometres in the southwestern corner. It bears many low cays in the southeast; others are scattered in the south, east, and north, and only a very small one in the west. These islets lie thus on the windward rather than the leeward side, a distribution similar to that observed in the Society Islands at Bora Bora, Mopelia and Scilly (Guilcher *et al*., 1969).

The submarine topography of Truk lagoon is irregular, with a series of depressions exceeding 40 fathoms (73 metres), generally not connected with passages across the barrier. Many coral knolls rise from the bottom. The interpretation of these features by Shepard (1970) is that the enclosed depressions result from solution of limestone during low sea-level phases, the knolls being either remnants of such karstification, or due to coral growth during the Holocene marine transgression, or both. These conclusions are similar to those from other coral areas as the Bahamas, Mayotte, Clipperton, etc. (see Chapter three, section 2).

Not far from Truk, Pohnpei (Ponape) Island and its barrier reef stand in striking contrast, with a large central island and a narrow lagoon with extensive mangroves along the volcanic shores. There are depressions in the lagoon which have clear connections with submarine valleys and passages (Tayama, 1952, Vol. 1, p. 37 and 238, and Vol. 2, Fig. 9; Shepard, 1970, p. 1910 and chart).

The almost-atoll of Gambier (Brousse *et al*., 1974; Davis, 1928, p. 376) (Fig. 95) has not been surveyed as closely as Truk; but an essential difference is that its barrier reef is drowned for at least half its length. Several possible explanations have been examined by Brousse *et al*., but none of them is definite. The first idea is that there has been tilting, so that the barrier reef subsided in the south, but this does not explain the drowned northwestern section, so that several local sinkings would perhaps be a better hypothesis. The volcano upon which the Gambier framework rests is dated at only 5 million years (Fig. 55), a surprisingly young age for such sinkings which normally would be expected later in the history of this structure. Another curious fact is the existence of the small Temoe Atoll 40 kilometres southeast of Gambier, and Portland Bank, 30 kilometres farther in the same direction, which was thought to be a shallow submarine bank, but remote sensing has

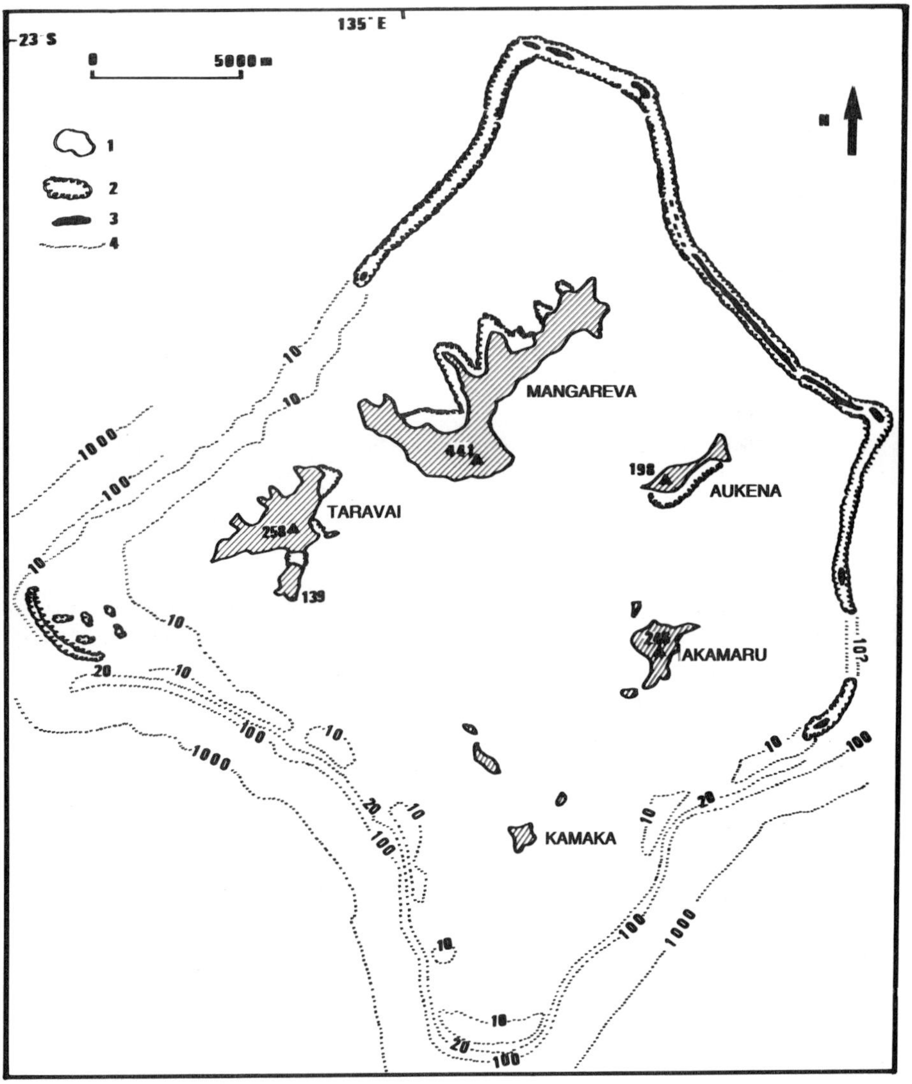

Fig. 95 The Gambier almost-atoll, Eastern Polynesia. After Brousse, Chevalier, Denizot and Salvat (1974), and the French Hydrographic Office. 1: high (volcanic) islands, with heights in metres; 2: barrier reef and main fringing reefs at sea-level; 3: low islands (motus) on barrier reef; 4: contours at 10, 20, 100 and 1000 metres below sea-level on the barrier (not shown in lagoon)

indicated that it is really another submerged atoll, under 12–13 metres of water (Pirazzoli, 1984). Atolls were not expected here, because in this direction the type of reefs should have been younger, not older than the Gambier almost-atoll. It thus seems that events have occurred here, which do not fit the general history of the Pitcairn–Mururoa–Hereheretue chain, and will require further investigation.

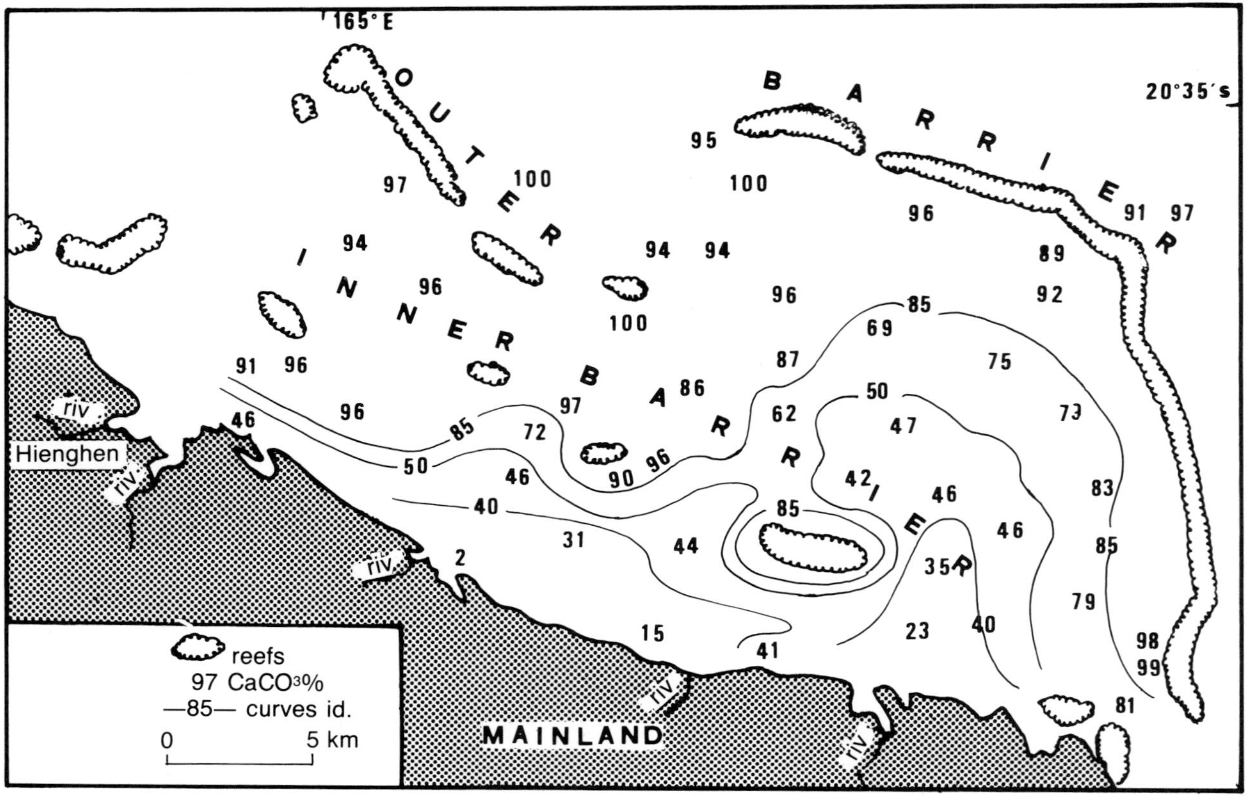

Fig. 96 Double barrier reef off northeastern New Caledonia, Melanesia. The sediments are almost completely calcareous in the outer lagoon, but strongly influenced by terrigenous supply in the inner lagoon. After Guilcher (1965), Singer Polignac Expedition

On the other hand, the Gambier lagoon depths, although not well known, are certainly very irregular. The lagoon has many coral knolls, especially in the north. The maximum depth which has been observed is 80 metres, between Mangareva and Akamaru: it is much deeper than all passages across the barrier, and thus seems to require, as at Truk, a karstic explanation.

Several almost-atolls exist in the Fiji Islands, east of Vanua Levu (Budd Reef), south of Viti Levu (Mbengha), and in the Lau Group (Exploring Isles, Ongea Levu, Yangasa Levu), interspersed with atolls and with 'normal' barrier reefs.

3.5 Double and multiple barriers

A general review of these structures was made by Guilcher (1976), and subsequently a similar structure has been described in the Philippines by Pichon (1977). For further discussion and new data see Coudray *et al.* (1985).

Although infrequent, double and multiple barrier reefs are not rare, as is sometimes said. They have been described from Mayotte in the Indian Ocean, Vanua Levu and Viti Levu in Fiji, several places around New Caledonia, off Queensland between Cape Melville and Lookout Point, North Bohol in the

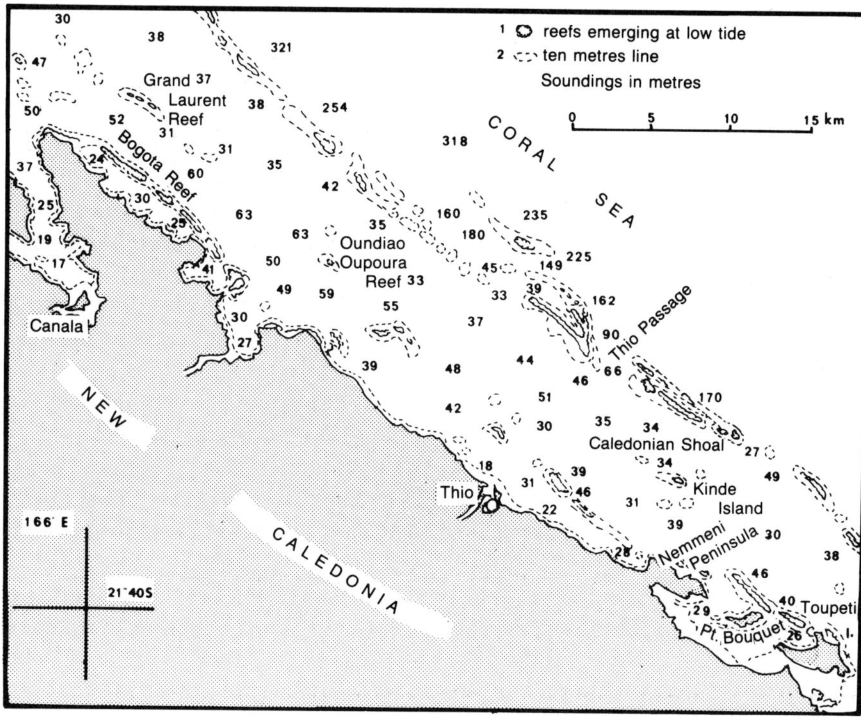

Fig. 97 Multiple barrier reef off Canala and Thio, New Caledonia. After French charts (from Guilcher, 1976)

Philippines, Wallis Island in Polynesia, and around the New Georgia Group in the Solomon Islands (Figs. 85, 96 and 97).

Since barrier reefs are a result of subsidence of the land, double or multiple barriers could be expected in places where subsidence has occurred in successive phases accompanied by faulting. As a matter of fact, faulting has been considered as the reason for the inner barrier reef at Mayotte by Guilcher *et al.* (1965a), the coral growing on the landward side of a fault scarp bordering an inner subsided area (Fig. 85). This explanation has been developed by Coudray *et al.* (1985), who extended it to account for the multiple barrier reefs found in parts of New Caledonia. Around New Georgia, tectonic effects have been demonstrated by Stoddart (1969c), since barrier reefs have been strongly uplifted above sea level (see Chapter three, section 7).

The problem, however, is to understand in which way corals were able to build barrier reefs inside a lagoon, where, as Davis (1928, p. 29) pointed out, the ecological conditions of growth are much poorer than along the outer barrier reef, washed by the oceanic swell. It is true that fringing reefs are common along inner shores of lagoons behind barriers; but these are only thin coatings, not massive as barrier reefs.

Guilcher (1976) has suggested to consider five factors able to permit unusually favourable conditions of coral growth in these lagoons: firstly, a large tidal range allowing oceanic swell to penetrate freely over the outer barrier at high spring tides; secondly, many wide passages across the outer barrier, which have the same result; thirdly, a large distance between the outer barrier and the mainland, resulting in a long fetch which favours strong waves generated by local winds and thus good water circulation; fourthly, a lagoon stretching in the direction of the dominant wind, again permitting strong wave action and water stirrings; and fifthly, strong currents entering the lagoon through passages, and favouring the growth of coral ridges.

The Mayotte double barrier reef can probably be explained largely by the first factor, because the tidal range reaches 3.50 metres at spring tides. In New Caledonia, the second and the fourth factors seem to be dominant, especially the fourth one, since the prevailing wind, blowing from the southeast along the axis of the lagoon, produces large waves. At North Bohol, Pichon (1977) suggested that strong tidal currents (i.e. the fifth factor) were the main cause. Off Queensland, where the tidal range is large and the lagoon wide, the first and third factors are relevant; at Wallis Island, the fifth and perhaps the first, and at Vanua Levu, the fifth (see for Vanua Levu Kuenen, 1950, Fig. p. 470).

Conversely, the absence of double barriers in the Society Islands seems to be due to a combination of a very small tidal range, narrow lagoons, and lack of wide passages on the windward side, except at Tahiti. None of the suggested possible explanations, however, appears to apply to Viti Levu; and around New Georgia the barrier pattern remains difficult to understand in its details; but the intense tectonic activity may have modified the ecological conditions for coral growth in that area since the reefs began to be built.

3.6. Sand cay reefs and low isles in lagoons

This subtype of coral reefs has been described more frequently and accurately than any other one perhaps. Fairbridge (1950), Stoddart and Steers (1977), Stoddart *et al.* (1978), Hopley (1982, p. 317–355), all concerning the Great Barrier Reef, may be selected among references since they refer to previous papers and give general views. As Steers (1929, p. 241) pointed out, 'the term Great Barrier Reef is, some ways, extremely misleading', since this huge area 'really consists of a number of reefs and reef patches cut by many channels' among which reefs with sand cays are widespread. Similar features have been described from Jamaica (Steers, 1940; Steers *et al.*, 1940), Indonesia (Kuenen, 1933; Umbgrove, 1947; Verstappen, 1953, 1954), off the northwestern and southwestern coasts of Madagascar (Guilcher, 1956; Salomon, 1980), off eastern Guadalcanal (Stoddart, 1969e), etc.

Sand cays are found at various stages of growth in lagoons behind barrier reefs, as in Queensland, and around New Caledonia where they have not been described so frequently as in Australia although they are quite numerous and fine. They occur on reef patches in shallow, wide lagoons where there is a long fetch in the direction of dominant winds, resulting in strong wave action. Substantial amounts of reef debris are thus produced, and swept up on to the reef flat to form surface accumulations. In most cases, these accumulations

Fig. 98 Several sand cays at different stages in development on reefs in the eastern lagoon of New Caledonia. All reefs face the southeasterly trade winds (from the left on the picture) (Photo A. Guilcher)

include ramparts of coarse material on the windward side, and islets (cays) of sandy sediments, more or less vegetated, on the leeward side. A 'lagoon', generally extremely shallow, occurs in between, sometimes with mangroves. Beach rock is frequent in the leeward cays.

On Fig. 98, different stages in sand cay development may be observed on New Caledonian reefs, with a simple tongue of loose sand on the reef in the foreground, two well-developed structures in the middle and far distances, and another incipient, circular structure in between.

The so-called Low Isles in the Great Barrier Reef environment (Fig. 99) are a variable structure, with modifications that have been investigated several times since it was first surveyed in 1929. Then there were three ramparts of coarse material on the windward side, extensive mangroves in the eastern part of the 'lagoon' (normally very shallow here), and a leeward sand cay with some beach rock. Subsequent surveys indicated that one or two other ramparts were added before 1945, but surveys in 1954 and 1973 showed the

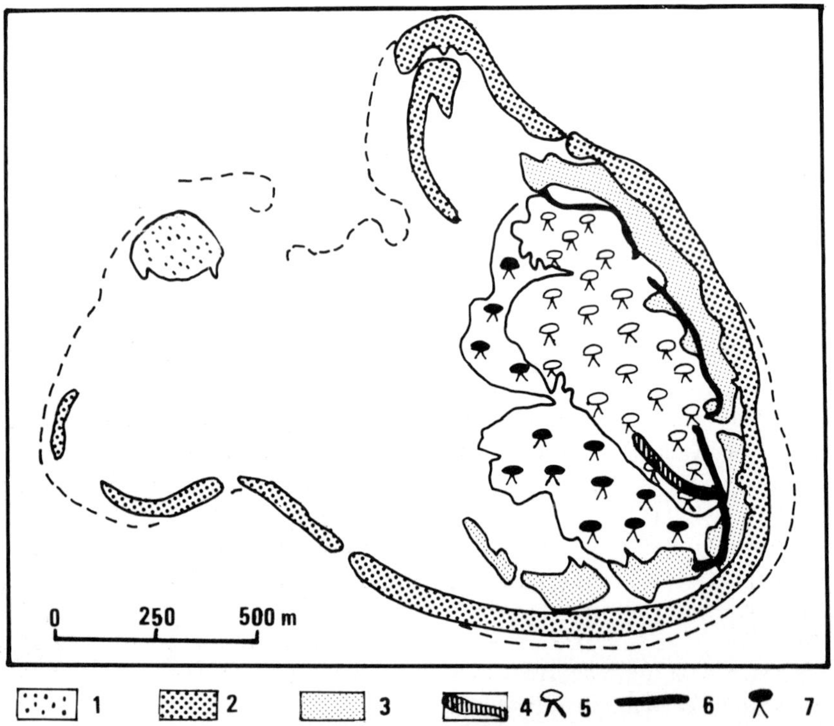

Fig. 99 Low Isles, Great Barrier Reef near Cairns at 16°S. After Stoddart *et al.* (1978), simplified. 1 to 5: situation in 1929. 1: sand cay; 2: outer shingle rampart; 3: inner shingle rampart; 4: older shingle ridge; 5: mangroves; 6: white shingle ridge in 1973, to which the structure has retreated, leaving only a few eastward remnants; 7: westward expansion of mangroves by 1973

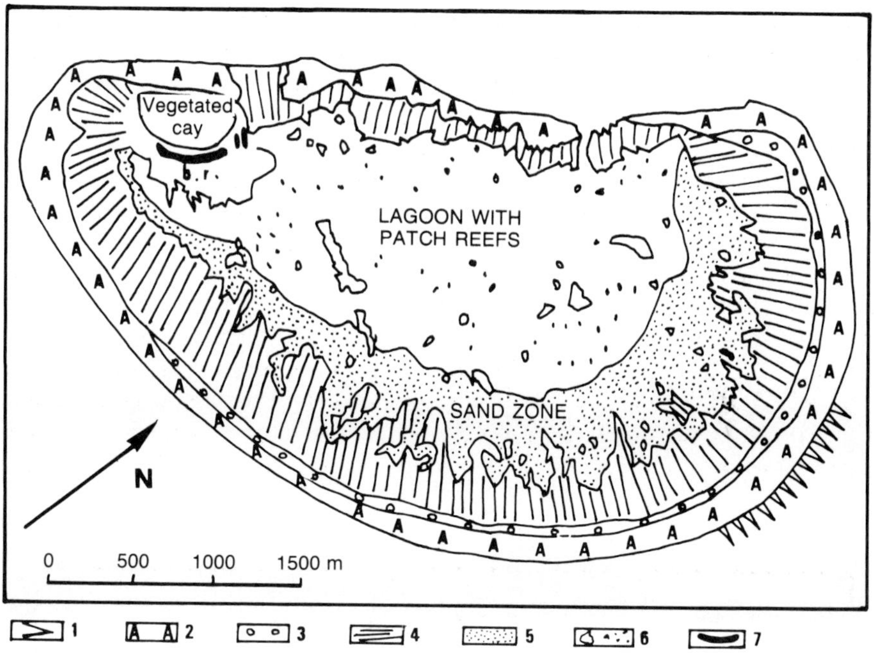

Fig. 100 Lady Musgrave Reef, at the southern end of the Great Barrier Reef (152°23′E, 23°54′S): a mature, large cay reef. After Orme *et al.* (1974). 1: spurs and grooves on windward side; 2: algal rim; 3: coral rubble (mainly Acropora debris); 4: radial lineations; 5: sand zone; 6: lagoon (generally less than 6 metres in depth) with many patch reefs; 7: beach rock along vegetated cay, which consists of sand and shingle

destruction of considerable parts of these outer frameworks, and progradation to the west of their remnants, presumably as the results of recurrent tropical cyclones. Yet the mangrove area has enlarged, and is thought to have impeded the migration of ramparts. The sand cay has not moved much, probably because it is partly anchored on beach rock, but the exposure of layers of beach rock on its southeastern shore testifies to slight reshaping. Evidence from the Low Isles thus indicates that these structures do not necessarily show continuous growth, but can be diminished in area or change in shape during stormy phases.

Also in the Great Barrier Reef area, Lady Musgrave Reef (Fig. 100) is a fully mature structure which, in fact, is not wholly typical since it contains a real lagoon a few metres deep, that is, a little deeper than usual: Lady Musgrave Island has been sometimes considered as an atoll. But this reef displays an excellent zonation, with, from outside, spurs and grooves, a calcareous algal rim, a narrow crown of coarse sediment, a beautifully striated zone, a sand zone, a lagoon with small coral heads, and, on the leeward side, a mature, vegetated cay.

Fig. 101 The Hiengu sand cay reef in Tuo Lagoon, east coast of New Caledonia, an example of a fully mature sand cay, facing the southeasterly trade winds (from the right of the picture) (Photo A. Guilcher)

The Hiengu sand cay on a patch reef near Tuo, in the eastern lagoon of New Caledonia (Fig. 101), is another example of a mature structure. It is covered by a dense broadleaf woodland, and occupies a larger part of the reef surface than is general. The windward side of the reef normally bears a recurved ridge of small blocks. The 'lagoon' between this and the sand cay is almost dry at lowest spring tides. Slabs of beach rock exposed on the windward shore of the cay indicate that a leeward migration has occurred, shown also by the bare sand tail prograded on the other shore.

Chivas *et al*. (1986) investigated the development of a coral cay, Lady Elliot Island, on the Great Barrier Reef, by means of radiocarbon dating over the last 3200 years. Accumulation rates were nearly uniform, and the cementation of shingle ridges by guano-derived phosphate has thus been measured accurately. Such dating should be applied to other coral cays.

In Indonesia, Verstappen (1954) has shown that changes in the outlines of sand cays and ramparts have been related to variations in strengths of alternating monsoons over periods of several years. In the Caribbean, hurricanes erode and may sweep away sand cays, probably even more than in other coral areas. Causes of cay instability have been investigated by Flood and Heatwole (1986), and Hopley (1982, p. 354; has remarked that 'variations in reef-top ages, morphology, and sea level history combined with

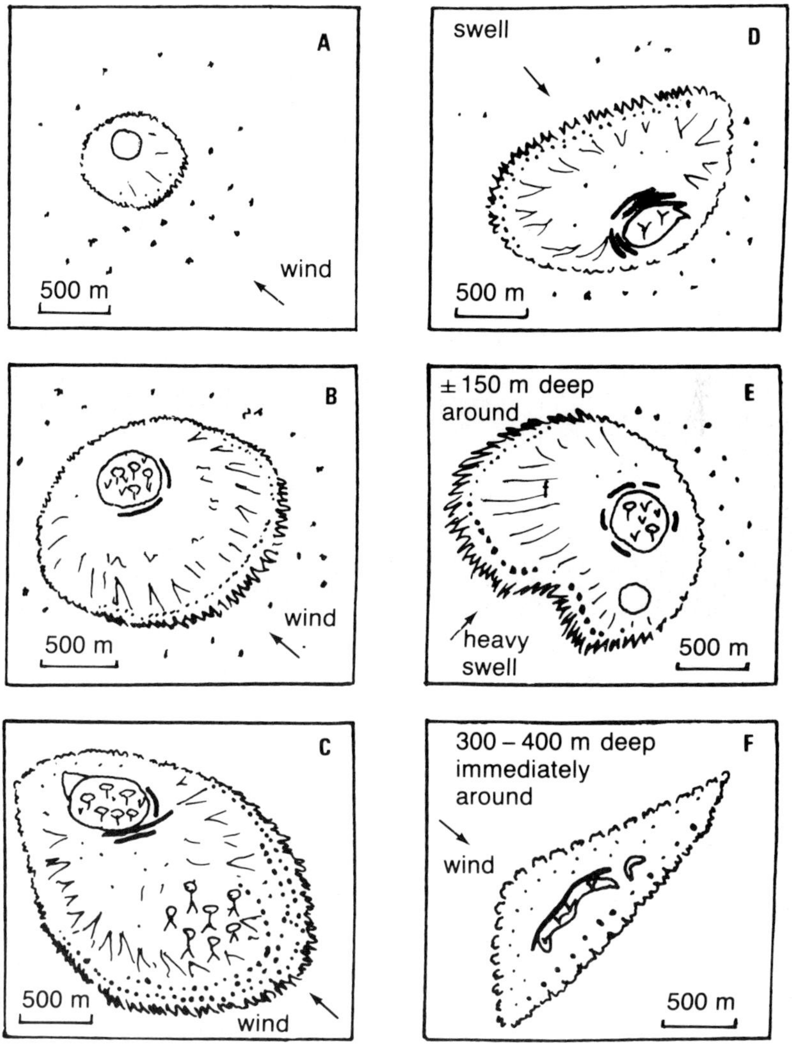

Fig. 102 Sand cay reefs. Evolution and subtypes. Surrounding depths: 15–25 metres in (a), (b), (c), and (d); deeper in (e) and (f). (a)–(b)–(c): evolution in lagoons behind barrier reefs in New Caledonia and Queensland. Subject to many variations in shape and size; can be strongly modified by hurricanes. Mangroves frequent in (c), in Queensland, behind shingle ramparts on windward side, but rare in New Caledonia. After various sources; (d): on the northwestern shelf of Madagascar, produced by the northwesterly swell; waves generated by the southeasterly trade winds are less effective because of the short fetch. After Guilcher (1956); (e) on the southwestern shelf of Madagascar. Heavy swell from the Southern Ocean, hence outer ridge of big blocks facing southwest. Two sand cays on the same reef, the southern one being submerged at high tide. After Salomon (1980); (f) sand cay capping the huge Marmar coral pinnacle on Farsan Bank, Saudi Arabia, Red Sea. Large blocks surprisingly located on leeward side. After Guilcher (1955). In each of the diagrams the heavy lines alongside cay show beach rock

differences in energy conditions and tidal ranges' have produced a large diversity, especially in the Great Barrier Reef province.

Taking account of these factors of differentiation, destruction and rebuilding, a general model of the growth of sand cay reefs is tentatively proposed here (Fig. 102), acknowledging that it may not apply to every case.

4. ATOLLS

There are 425 atolls recorded in the world (Stoddart, 1965); a few more if submerged atolls and some of the bank atolls (this chapter, section 2) are taken into account. The archipelago with the largest number of atolls is the

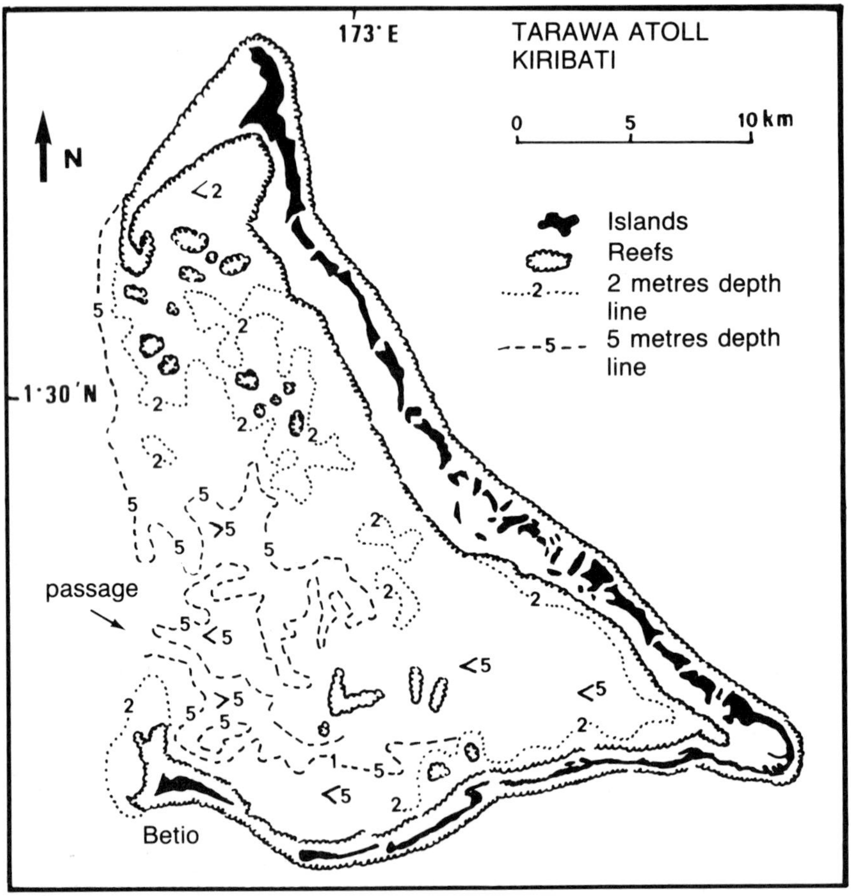

Fig. 103 Tarawa Atoll, Kiribati, Micronesia. An asymmetrical atoll, without a rim on the leeward side (compare Rose Atoll, Samoa, Fig. 58), and with a very shallow lagoon. Other atolls of the Kiribati group are also asymmetric, as Butaritari, Nonouti, Maiana. Hoas and teiruas (see text, section 4.6) exist on the northern side between islands

Tuamotus (76). The largest atoll is Kwajalein in the Marshall Islands (120 × 32 kilometres), followed by Rangiroa in the Tuamotus (79 × 34 kilometres), and Tijger, south of Celebes (72 × 36 kilometres). Twenty-seven atolls are recorded in the Caribbean (Stoddart, 1965), but perhaps only 10 deserve that name (Milliman, 1973). The problem of the existence of true atolls in the Red Sea is discussed in section 5 of this chapter (ridge reefs). The most comprehensive study of atolls is by Wiens (1962).

4.1 Asymmetry

Atolls are generally more or less asymmetrical, with differences between the windward and leeward sides. Such differences are well known at Bikini and nearby atolls (Emery *et al.*, 1954), especially in the distribution and slope of the spur and groove systems (Chapter two, section 1.2). In the Society and Tuamotu Islands, the low isles on atoll rims are principally or exclusively on the windward side, where sediments are thrown up by waves, and such a distribution is considerably exaggerated in Kiribati atolls such as Tarawa (Fig. 103), Apaiang, Maiana, Abemama, and Butaritari (Guilcher, 1967b), in Caribbean atolls such as the Turneffe Islands (Stoddart, 1962a), and even more in the atolls off the shelf of Nicaragua (Milliman, 1969; Milliman and Supko, 1968), where the leeward side is almost absent, so that the atoll consists only of an arc facing east (Fig. 104). Such atolls are transitional towards the arcuate reefs forming the Great Barrier Reef (Fig. 88). By contrast, low isles are almost continuous around the atoll rim at Diego Garcia in the Indian Ocean (Stoddart and Taylor, 1971, p. 27).

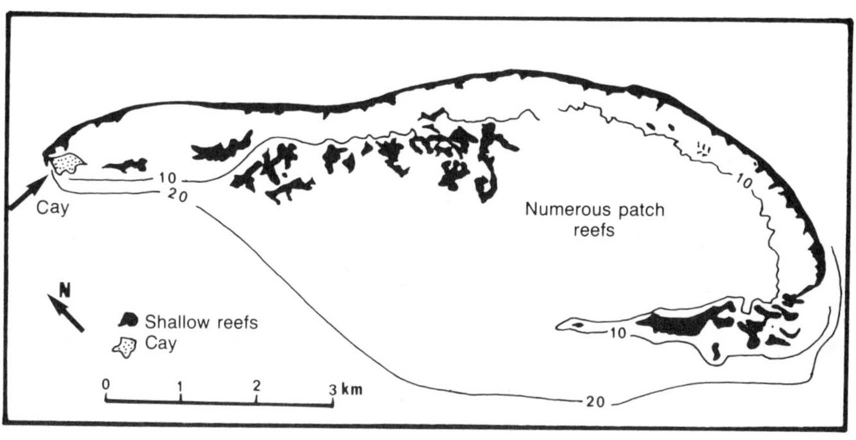

Fig. 104 Roncador Bank, a Caribbean atoll off Nicaragua, with wide opening on its leeward side. Depth contours in metres. After Milliman (1969)

Fig. 105 Mataiva atoll, Tuamotu Islands. Western rim. Pink *Porolithon* forms outer edge with spurs and grooves. Reef flat with jagged (bioeroded) remnants of a Holocene ledge built to a slightly higher level (Photo A. Guilcher)

4.2 Atoll history kept in rim structure

Atoll rims often retain remnants of their Pleistocene or early Holocene structure, and such features can occupy more space than present growths. This is particularly true for the Tuamotus, where Agassiz (1903a) has given descriptions of what he called the 'old ledge' of the atolls (Fig. 105); also at Mopelia in the Society Islands (Fig. 106); in some places, even Tertiary (Mio-Pliocene) remnants can still be seen at the surface, as at Mataiva Atoll (Fig. 107; see section 4.8 of this chapter for details).

Fossil parts of the rims are sometimes located in places which raise problems for explaining atoll evolution. Such is the case in several atolls of the northwestern part of the Tuamotus (Rangiroa, Kaukura, Anaa, Tikehau: see Stoddart, 1969b; Guilcher *et al*., 1969), where Pleistocene corals, highly dissected by bioerosion, stand as pinnacles several metres high (the so-called *feos*: Fig. 56), only on the south side of the rims. The explanation put forward by Pirazzoli and Montaggioni (1985) is, as already said in Chapter three, section 6, that this local uplift of the fossil rim would be related to the moat and arch effect in the tectonic history of the Society chain formation.

Even apart from atolls standing several tens of metres above present sea-level (see section 4.13), there are atolls with a central lagoon that consist exclusively of old corals in their upper rims. Such is the case at Aldabra,

Fig. 106 Mopelia Atoll, Society Islands (located on Fig. 54). Mushroom rocks as residuals of dissection of slightly emerged Holocene (3450 BP) reef (Photo A. Guilcher)

Fig. 107 Mataiva Atoll, Tuamotu Islands (located on Fig. 34), north coast. Oldest (Tertiary) reef outcrop, highly dolomitized and recrystallized (Photo A. Guilcher)

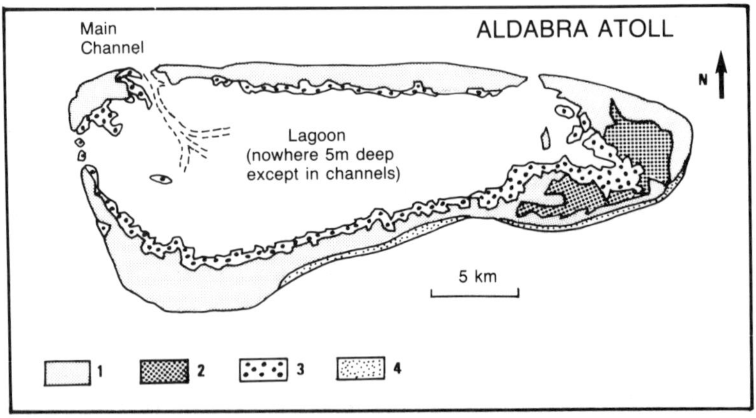

Fig. 108 Aldabra Atoll, Western Indian Ocean. A slightly emerged atoll. Simplified from Stoddart *et al.* (1971). 1: champignon (jagged fossil coral, with several metres relief in places); 2: platin (smoother fossil coral, also emerged); 3: mangroves along lagoon shores; 4: beach and dune sand

southwestern Indian Ocean, which has been investigated by Stoddart *et al.* (1971) (Fig. 108). Its rim, cut by three passages connecting a lagoon 6 to 10 kilometres wide with the open sea, consists of Pleistocene corals in position of growth, belonging to two successive units of different composition and crystallization, with a maximum height of 7.5 metres; these emerged fossil corals have been intensely dissected on the lagoon shore to form low cliffs with deep notches preceded by jagged pinnacles. According to Trudgill (1976), cliff retreat is due primarily to bioerosion, accompanied by solution, salt weathering, and abrasion. These cliffs are backed by rasp-like surfaces called 'champignon', while smoother surfaces with lower relief form the 'platin'. Small unconsolidated dunes occur, especially on the windward coast. The lagoon is less than 5 metres deep, except in channels that open towards the passages, which exceed 20 metres in depth, and are scoured by tidal currents (the tidal range exceeds 3 metres at spring tides). Mangroves grow on the lagoon shores. Aldabra is thus an excellent example of an atoll combining fossil and modern geomorphological features, the fossil ones playing an important role in a still living structure.

Situated in the Mozambique Channel, Europa Atoll, described by Battistini (1966), resembles Aldabra very much, but is smaller (6 × 6 kilometres). It is also slightly emerged, and probably of Upper Pleistocene age, jagged by karstification, with a lagoon shallower than at Aldabra in which grow extensive mangroves of various species.

4.3 Lagoon sediments

Information on atoll lagoon sediments has been given at Bikini, Enewetak, Rongerik, and Rongelap in the Marshall Islands by Emery *et al.* (1954); at

Johnston, an isolated atoll-like feature between Hawaii and the Line Islands by Emery (1956b); at Kapingamarangi in the Carolines by McKee *et al.* (1959); at Mopelia in the Society Islands by Guilcher *et al.* (1969); at Kure and Midway in Hawaii by Gross *et al.* (1969); in the Caribbean by Milliman (1973), and by Guilcher (1967a) for several areas.

In contrast with sediments in lagoons between barrier reefs and high volcanic or sedimentary hinterlands, sediments in atoll lagoons are entirely calcareous, except for a few diatoms and spicules of silica embedded in sponges, and sometimes traces of aeolian dust. The sources of supply are coral particles of various sizes, derived from coral knolls living in the lagoon (see section 4.5), and partly from grazing by fishes (Chapter two, section 4); among algae, *Halimeda* is a major source, as well as foraminifera which live in most lagoons; and various shell fragments from echinoderms, gastropods,

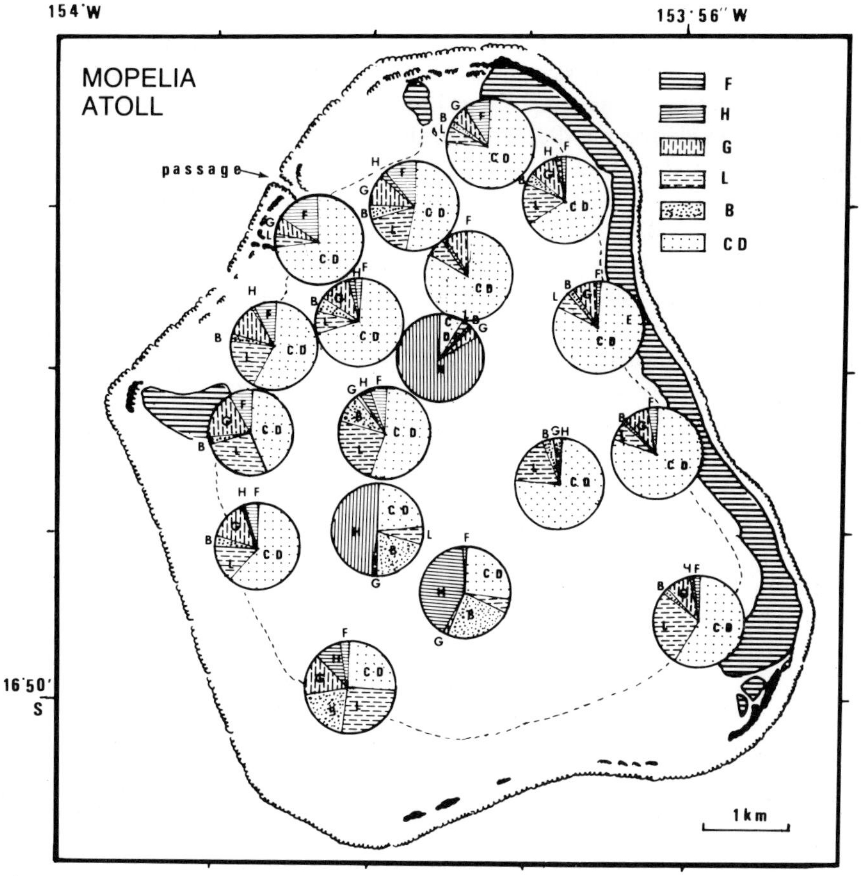

Fig. 109 Mopelia Atoll, Society Islands (located on Fig. 54). Sedimentary composition of lagoon deposits. After Berthois in Guilcher *et al.* (1969). F: Foraminifera; H: Halimeda; G: Gastropoda; L: Lamellibranchiata; B: Bryozoa; CD: coral debris

lamellibranchs, crustaceans, sponges, and other organisms, provide skeletal deposits. These sediments can either settle *in situ*, or be transported across lagoons by currents (see section 4.7). Trituration of bottom sand by holothurians, which live in large quantities in lagoons, does not seem to modify the particles to some extent (Cloud, 1959, and others). On precipitation of oolitic limestone, mostly in the Bahamas and the Persian Gulf, see this chapter, section 6.

Halimeda generally plays a considerable part in this sedimentation, particularly in deep water (at 45–50 metres depth at Bikini, Rongerik, Rongelap, and in the deepest parts at Mopelia where they form 42 to 74 per cent of the sediment: see Fig. 109), or a ring-like pattern around the central area (at Enewetak and Kapingamarangi). By contrast, *Halimeda* debris is rare at any depth at Johnston, where coralline algae play a large part. Foraminifera are mostly represented in beaches and nearby shallow waters (up to 85 per cent of the sand grains in beach sands at Rongerik; more scattered, however, at Mopelia). Foraminifera found in deeper water at Kapingamarangi and Enewetak are of different species to those settling in shallower areas. Fine coral debris occurs principally near the rim of the four Marshall atolls, decreasing towards the middle, and the distribution of gastropod and lamellibranch shells is the same at Mopelia. Coarse coral debris is found at all depths, even in the central parts where it settles around the coral knolls, at the foot of which talus is visible at Bikini and Mopelia.

At Kure and Midway, the world's northernmost atolls, the sediments in lagoons are, in decreasing order of abundance, foraminifera, coralline algae, shell fragments, coral fragments, and *Halimeda*. The scarcity of *Halimeda* may be due to the high latitude (28°25′, 28°13′), but as it is also rare at Johnston (17°), it may be a special feature of the Hawaiian region. In Caribbean atolls, *Halimeda* is often abundant in sediments, as at Serrana, Roncador, and Alacran (Milliman, 1973).

4.4. Mangroves

Mangroves are not so frequent in atoll lagoons as on the inner shores of fringing reefs and in 'low wooded islands' in Queensland, Madagascar, the Red Sea, the Caribbean, and Indonesia. They were not reported in the Marshall Islands (Emery *et al*., 1954), and appear to be completely absent from the Society and Tuamotu Islands, in spite to attempts to introduce *Rhizophora* to the Moorea lagoon in 1965. They were not mentioned in the four atolls described by Milliman (1969) off Nicaragua. On the contrary, *Rhizophora* thrives in sheltered parts of the Abemama and Tarawa lagoons in Kiribati, where it grows vigorously on calcareous sands and muds (Guilcher, 1967) (Fig. 110). Mangroves are also present in the Carolines, for example at Losap Atoll (Tayama, 1952, vol. of plates p. 102), at Aldabra and Europa Atolls in the Indian Ocean, as said previously, and in two of the three atoll lagoons near Belize, Caribbean (Stoddart, 1962a).

Fig. 110 Abemama Atoll, Kiribati, Micronesia. Mangroves (Rhizophora) growing at Binoinano, a sheltered place in the lagoon, on medium sand. Low tide (Photo A. Guilcher)

4.5 Pinnacles or knolls in lagoons

The floors of many lagoons bear numerous pinnacles or knolls, which rise from various depths. Fig. 111 gives an example from Bellingshausen Atoll in the Society Islands, which has no passage and is fed with surface oceanic water only by overwash: in spite of this, coral growth appears to be excellent. Pinnacles rising to the surface are a common occurrence in the Tuamotus, for example at Arutua (Fig. 112), Manihi, Ahe, Tikehau, Mururoa. In the Indian Ocean, Diego Garcia (Stoddart and Taylor, 1971, figs. 13 and 14) is another striking example. The pinnacles generally have living corals on their surfaces.

However, a host of pinnacles in lagoons is not a universal feature. For example, they appear to be rare at Scilly in the Society Islands, where no knoll is visible at depths shallower than 12 metres (Guilcher *et al*., 1969, p. 39). There is apparently no relation between the occurrence or number of pinnacles and the degree of connection between the lagoon and the open sea, and this is true also in lagoons behind barrier reefs: Bora Bora lagoon has good connections with the ocean but no knolls, while Tupai, Tetiaroa, Bellingshausen (Society) and Kaukura (Tuamotus) have knolls and no deep passage; Rangiroa (Tuamotus) has knolls and two deep passages; and the general situation is the same at Bikini, Rongerik, Enewetak and Rongelap in the Marshall Islands.

Fig. 111 Lagoon of Bellingshausen Atoll, Society Islands (located on Fig. 54), full of coral knolls although without deep passage (see Fig. 117) (Photo A. Guilcher)

Fig. 112 Arutua Atoll, Tuamotu Islands (located on Fig. 34). Many pinnacles in lagoon, one bearing an islet. In foreground, a spit has grown to partly enclose a hoa which thus becomes a tairua (see Fig. 115) (Photo A. Guilcher)

Since pinnacles do not seem especially related to present conditions of vigorous coral growth, another possibility is that they are remnants of karstification during low sea-level phases, a hypothesis favoured by supporters of the karstic saucer theory (Chapter three, section 5). But pinnacles can be of composite origin: their core can be Pleistocene, and their surface or upper part Holocene; some can be old and others recent. Only the higher ones, rising to present sea-level, can be taken, if they are Pleistocene, as evidence for an antecedent platform beneath the present atoll lagoon. The significance of the pinnacles will depend on numerous sections in many places, and radiometric dates. A few sections and dates will not be decisive. If adequate investigations are carried out in this way, the clue—or one of the clues—of atoll genesis may be there.

4.6. Ava, hoa, tairua/passage, overflow channel, temporarily closed channel

In the Tuamotus, the communication between atoll lagoons and the ocean occurs in two ways, recognized in Paumotu terminology (Vallaux, 1955; Guilcher, 1967b; Chevalier *et al.*, 1968; Guilcher *et al.*, 1969; Chevalier, 1972). There are a few passages several metres deep, which are suitable for navigation, called *ava*, which have been discussed in Chapter three, section 2 (Fig. 34). In addition, there are many (several thousand in the Tuamotu archipelago) very shallow passages, scattered around the atolls between the

Fig. 113 Rangiroa Atoll, Tuamotu Islands (located on Fig. 34). Typical succession of motus and hoas on the western part of the atoll rim (Photo A. Guilcher)

Fig. 114 Maiana Atoll, Kiribati, Micronesia, southwest corner. Hoas, with deltas of sediment washed into the lagoon as over barrier reefs in the Society Islands (cf. Figs. 91 and 92) (Photo A. Guilcher)

motu(*s*) (islets) but better defined than the low flat areas through which swash penetrates in most cases into the Marshall lagoons. These are known as *hoa*(*s*), a word which can be translated as overflow or overwash channel; with the intervening elongated motus, they result in the typical Tuamotu landscape (Fig. 113). They have been supposed by Chevalier *et al*. (1968) to result from joints formed by compaction of the reef, and to have been subsequently widened by erosion. They act efficiently as lagoon feeders in water. Series of hoas with intervening motus have also been observed in the Kiribati Group at Tarawa, Abemama and Maiana (Guilcher, 1967b) (Fig. 114), and even at Majuro in the Marshall Islands (Newell and Bloom, 1970, p. 1885), an archipelago where they seem, however, much more exceptional than in the Tuamotus, while the avas are more numerous.

But, as the Tuamotu lagoons are often wide, the fetch is large, and large waves are generated when the wind is strong. These lagoon waves are able to rework the sandy beaches on the inner side of the motus, and build long sand spits across the inner ends of the hoas during periods of moderate oceanic swell, hence lack of currents through the hoas (Guilcher, 1966; Harmelin-Vivien, 1985, p. 232 and 254) (Figs. 115 and 116). Agassiz (1903a) already observed these spits. Ponds are thus formed in the hoas, which become *tairua*(*s*)

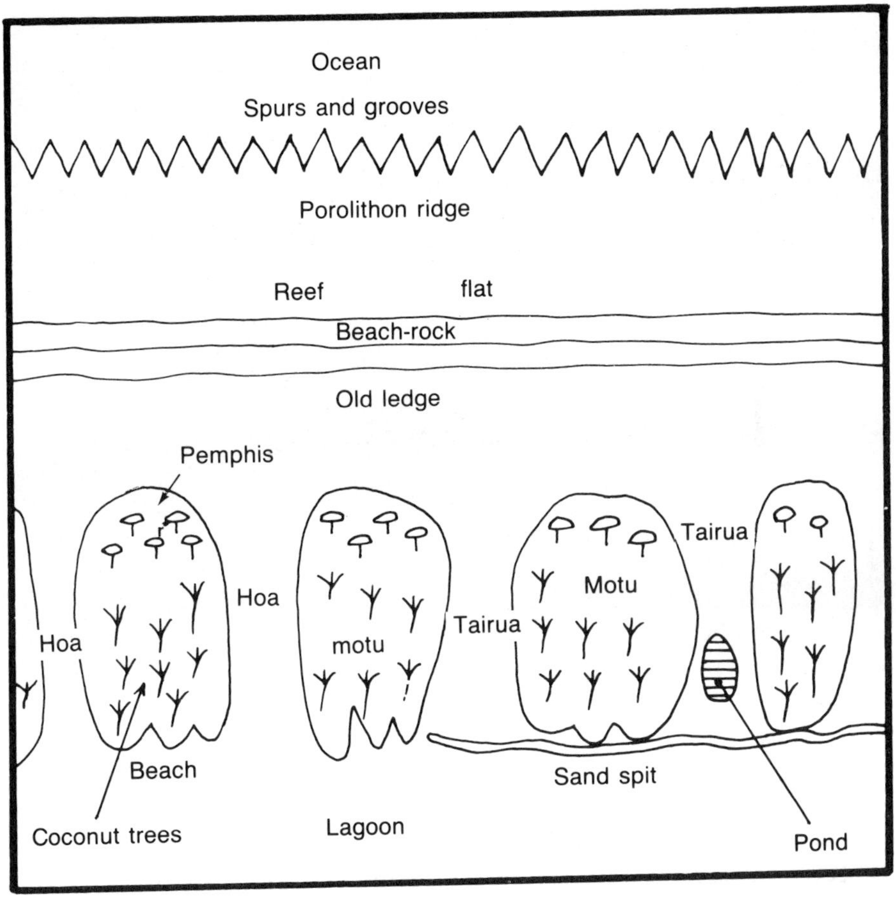

Fig. 115 Elements of an atoll rim in the Tuamotus. After Guilcher (1966)

and are more or less persistent. Some have been used for aquaculture at Rangiroa Atoll. They also exist in the Bora Bora barrier (Guilcher *et al.*, 1969, p. 46–47; Pirazzoli, 1985, p. 44), and at Tarawa, where open hoas are, however, more frequent (Guilcher, 1967b) (Fig. 103).

Avas, as at Mopelia and Rangiroa, are generally deeper in their central part than at the lagoon entrance, which remains shallow because of the decrease of scour by tidal currents.

A number of atolls in the Tuamotu and the Society Islands have no ava, so that the connection in surface with the ocean is possible only through hoas, or over the reef flat without hoas, the second case occurring mainly in the Society Islands as at Scilly, Tupai, and Bellingshausen (Figs. 117 and 118). An extreme situation has been analysed (Chevalier and Salvat, 1976) at Taiaro in the Tuamotus, an atoll 4.5 kilometres long and 3 kilometres wide (Fig. 119), which, like some others in the same group, is a closed atoll, the rim consisting

Fig. 116 Apataki Atoll, Tuamotu Islands (located on Fig. 34). Motus, hoas, and tairuas closed by the formation of sand spits on lagoonside (Photo A. Guilcher)

Fig. 117 Bellingshausen Atoll, Society Islands, an atoll without an ava (Photo A. Guilcher)

Fig. 118 Tupai Atoll, Society Islands (located on Fig. 54), an atoll without an ava but with more hoas than Bellingshausen (Photo A. Guilcher)

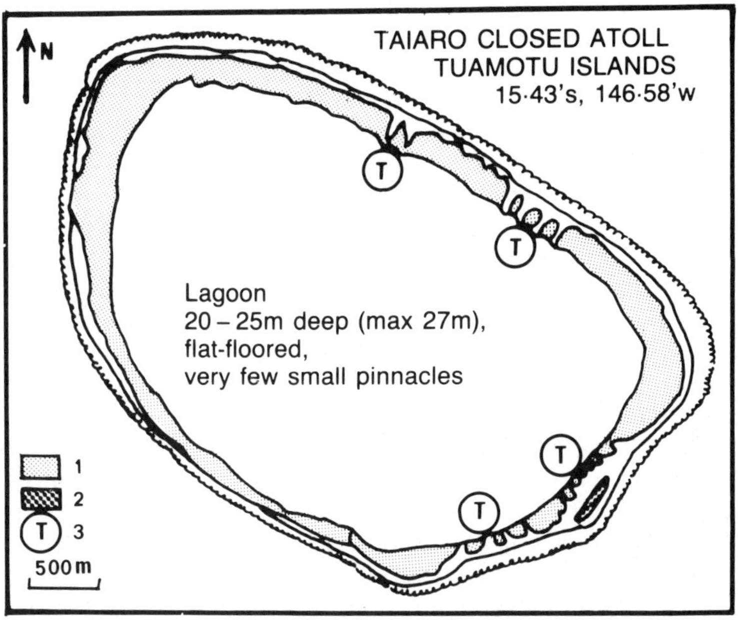

Fig. 119 Taiaro closed atoll in the Tuamotu Islands. After Chevalier and Salvat (1976). 1: terrestrial vegetation; 2: pond; 3: groups of tairuas (inactive except during exceptionally strong surf)

of a continuous circular island, built during a Holocene (1000–2000 BP) phase of sea-level 1.5 to 2 metres higher than now. It is breached by 18 tairuas, i.e. hoas which were open when the sea-level was higher. The tairuas act as hoas only in very severe storms. The level of the lagoon lies 0.60 metres below the open sea at high tide, and the water salinity is 43.5 to 44 per 1000, compared with 36 per 1000 in the ocean. The result is that the fauna in the lagoon has been considerably impoverished in the nearshore area, with only a single species of hermatypic coral (one *Porites*) and one echinoderm. Various shells, however, live in the lagoon, but it bears no important pinnacle, a fact not favourable to the karstic hypothesis of the evolution of atolls, since pinnacles should exist as residuals of solution of a former open reef platform. Takapoto, also in the Tuamotus (Salvat, 1979; Chevalier *et al*., 1979; Salvat and Richard, 1985) is not so closed as Taiaro: it has a few open hoas, and the lagoon has more than 400 pinnacles of various sizes with living corals on their surfaces.

4.7. Water circulation in lagoons and percolation at depth through the atolls

Lagoon circulation has been studied in several places, notably by Von Arx (1954) at Bikini and Rongelap in the Marshall Islands (see also Wiens, 1962). As expected, the main factor is the trade winds, which force swell into the lagoon over the windward, eastern side. The water in excess escapes partly over the leeward side, but some comes back to the east along the bottom, so that there is a division into rotating compartments. In addition, tidal currents run to and fro in the deep passages. The circulation is more marked when trade winds are strong, in the winter months of the northern hemisphere; it slackens in summer, when the doldrums settle near the two atolls.

Other observations have been made in the Society and the Tuamotu Islands (Guilcher, 1966; Guilcher *et al*., 1969, chapter 4) at the surface and on atoll rims, but not at depth. Except for the tidal currents which alternate through the avas, as in the Marshalls, the processes are different here. In addition to the trade winds which drive water in lagoons from the east, there is occasional or even frequent heavy swell from the south, generated in the roaring forties. This swell (Fig. 120) piles water into the lagoons freely, because islands are rare or absent on the southern side, being mostly located in the east, in the Society Islands at least. During such periods in 1963, the sea-level in Mopelia lagoon rose 40 to 50 centimetres above normal; the alternating tidal currents in the single deep passage were replaced by a strong and continuous flow out over the leeward side. On Scilly Atoll, such events have been frequent enough to create a morphology of crescentic accumulations of sand on the leeward rim, with convex sides that face the lagoon and not the ocean (Fig. 11). The surface circulation is the same at Mataiva Atoll and Tikehau Atoll in the Tuamotus, according to Delesalle (1985, p. 281), Delesalle *et al*) 1985, p. 36), and Harmelin-Vivien (1985, p. 221): oceanic water enters through the hoas in the south and east, and flows out through the ava in the northwest or

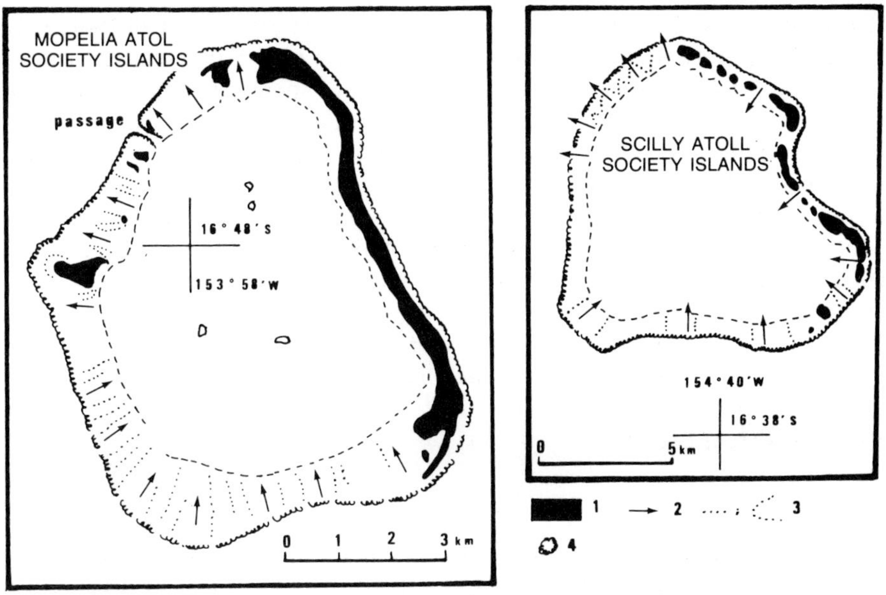

Fig. 120 Water surface circulation in the Mopelia and Scilly Atolls, Society Islands. After Guilcher (1966). 1: islands (motus); 2: inflowing and outflowing currents, generated by trade wind (easterly) swell or southerly swell; 3: sand stripes on atoll rim, convex towards lagoon on northwestern side (see also Fig. 11 for this feature). One passage (ava) on Mopelia, none on Scilly. Motus concentrated on eastern side. Pinnacles in Mopelia lagoon, none at Scilly

west. A similar pattern has been observed at Kaukura in the Tuamotus, but as that atoll has no ava, the water goes in and out only through the hoas, thus flowing lagoonwards or oceanwards according to their location (Guilcher *et al.*, 1969, plates 9 and 10).

In addition to this surface circulation, percolation certainly occurs at depth across the atolls, since they contain many voids, which may be original cavities or holes formed by karstification during low sea-level phases. The existence of a freshwater lens, called the Ghyben–Herzberg lens, at a shallow depth beneath the islets of the rim, has been known for a long time (Wiens, 1962, chapter 4), but it is now known that early descriptions of this lens are oversimplified (Oberdorfer and Buddemeier, 1985), although it certainly exists. At a lower level, oceanic water circulates through the karstified limestone. Studies carried out at Rangiroa and Mururoa in the Tuamotus (Michel, 1969; Michel *et al.*, 1971) have shown that the content of organic matter in the lagoon water is considerably higher than in the upper oceanic water surrounding the atolls: at Mururoa the biomass is more than ten times larger inside than outside. Ecologically, these Polynesian atolls thus appear as 'oases' scattered in the desert ocean surface. According to Sargent and Austin

(1954), the situation is the same at Rongelap Atoll in the Marshall Islands, in which 'the productivity is considerably higher than that of the adjacent waters or any other open marine areas'; these authors speak of 'flourishing coral atolls in an empty ocean'. Rougerie and Wauthy (1986, and in press) suggested that these contrasts result from an 'endo-upwelling' of deep oceanic water, richer in nutrients than the surface water, flowing across the inner, lower parts of the atolls. The role of this endo-upwelling process varies from one atoll to another: it seems rather weak at Taiaro Atoll (Chapter four, section 4.6), while at Takapoto, a half-closed atoll (same section), it seems stronger (Rougerie, 1979). At Enewetak in the Marshall Islands, it has been shown that effective permeability increases with increasing depth (Buddemeier and Holladay, 1977), a fact which may help to understand the deep percolation.

4.8. Reticulated atolls

Mataiva Atoll, 10 kilometres long and 5 kilometres wide (Figs. 121 and 122), at the westernmost end of the Tuamotus, has a very fine reticulated or honeycombed lagoon. This lagoon 'is made of about 70 pools of varying sizes (100 m to over 2 km), with an average depth of 8 m (down to 30 m). The shallow "ridges" which separate these pools are under 0.1–0.8 m of water and form a network more than 200 km long' (Delesalle, 1985). In a few

Fig. 121 Mataiva reticulated atoll, Tuamotu Islands (located on Fig. 34), as seen from the northeast at an altitude of 10,000 metres (Photo A. Guilcher)

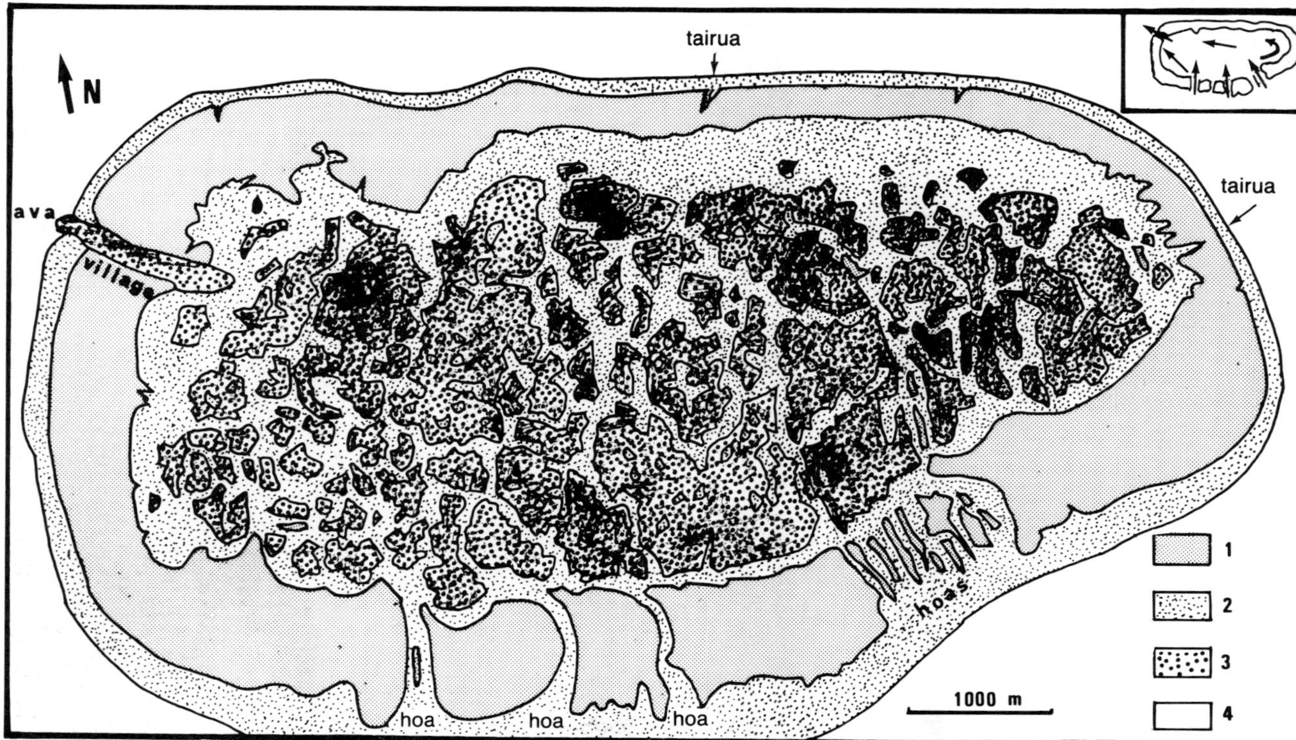

Fig. 122 Mataiva Atoll, from various sources. 1: emerged land (peripheral motus, and some tops of inner ridges, particularly in the east: see Fig. 123). 2: inner ridges, slightly submerged; beaches; hoas; outer reef. 3: deeper stretches of water in lagoon and in ava (passage). 4: ocean. Inset map: water circulation.

places, these ridges rise 20–30 cm above sea-level (Fig. 123), and are made of fossil corals of Middle Holocene age in position of growth. This strange pattern, and the rich phosphates which exist inside the cells, are though to derive from a complicated evolution. A reef platform, which outcrops on the north coast of the rim as a strongly dolomitized reef (Fig. 107), and dips southward below the lagoon, developed probably in the Mio-Pliocene. This platform was karstified during period(s) of emergence, possibly in the Late Pliocene, to produce the honeycombed pattern. The origin of the phosphates remains uncertain as in other atolls: they could have formed from bird excrements and bones, or from marine organic matter, or from drifted volcanic material, or from endo-upwelling according to Rougerie and Wauthy (in press). The last episode was a slight emergence after a Holocene maximum sea level, which appears not only in some lagood ridges (Fig. 123), but also in jagged upper parts of the southwestern outer rim (Fig. 105). There are two or three distinct generations of beach rock in some of the outer beaches, as on the east coast (Fig. 28). In the course of this evolution, tropical karstification appears to have occurred. But it is not clear why several uplifts and lowerings would have occurred at Mataiva, and apparently none in the other near-by Tuamotus. There are some typical hoas and tairuas (Figs. 124, 125, and 126).

The lagoon behind the barrier reef between Raiatea and Tahaa in the Society Islands displays a similar honeycombed pattern over a much smaller

Fig. 123 Mataiva Atoll. Emerged (fossil) Holocene coral ridge in the eastern part of the lagoon (Photo A. Guilcher)

Fig. 124 Mataiva Atoll, southern rim. Hitirari Hoa, looking towards the ocean. The water depth is less than a metre (Photo A. Guilcher)

Figure 125 Mataiva Atoll, northeastern rim. Wide Papiro Hoa, with in the foreground a large truncated coral colony of Holocene age in position of growth. Ocean in the distance (Photo A. Guilcher)

Fig. 126 Mataiva Atoll. A tairua (former hoa) in the east, now completely cut off from the ocean and invaded by vegetation (Photo A. Guilcher)

area. Hao Atoll in the Tuamotus also has a more or less similar structure in its lagoon, and Fakaofo in Tokelau seems to have too (Fig. 2 in Balazs, 1983). Canton Atoll in the Phoenix Islands, described by Smith and Henderson (1978), has a reticulated framework resembling Mataiva, but not so complete: the ridge network is less continuous and the cells have more interconnections. These ridges are covered near the surface principally by branching corals, but at lower levels the coral growth is poor or nil, because the atoll receives oceanic water only through one passage: there is no real hoa on the windward side, and the southeastern end is a cul-de-sac. No information is available on the internal structure of the ridges, but it is possible that karstification has occurred, although that hypothesis has not been put forward.

At first sight, it would seem that Farquhar Atoll in the western Indian Ocean (Battistini, 1977; Battistini and Jouannic, 1979) is of the same type as Mataiva and Canton: its lagoon contains more than twenty ridges, and in the southern and northern corners enclosed subsidiary lagoons occur. But the ridges are largely sandy, and always rectilinear and subparallel, and they do not enclose a set of cells as at Mataiva. The lagoon depth is less than 15 metres. The oldest parts of the atoll are Holocene (3000–3600 BP) and have slightly emerged (2 metres). It may be that Farquhar would not have been a true atoll, but a bank reef until quite recent times, with initially separate parallel sand stripes.

Christmas Atoll in the Line Islands (Valencia, 1977) is more truly reticulated than Farquhar, but it owes this aspect to some 500 hypersaline shallow lakes and lakelets, so that the lagoon proper is restricted to the northwestern corner. Christmas is certainly a very old structure (Chapter three, section 6), and could be called a dying atoll, like Jarvis Atoll, 420 kilometres to the southwest, which also has a filled lagoon floor, with evaporite deposits. It must be remembered that the Line and Phoenix Islands lie in the dry zone of the Equatorial Pacific.

In conclusion, reticulated patterns in atoll lagoons seem to be of diverse origin, although the most typical, at Mataiva and Canton, might result from more or less similar evolutions.

4.9. Drowned atolls

Drowned atolls are more numerous than reticulated atolls, and their history has been generally simpler.

Saya de Malha in the Indian Ocean has been investigated by Russians with a research submarine (Fedorov *et al*., 1980), and is typical. It forms a ring approximately 40 kilometres in diameter, with a central depression 70 metres deep in the north and 140 metres in the south, a peripheral rim only 8 metres deep at its top with depressions which look to be former passages. This rim still bears many living colonies of *Acropora*, *Pocillopora*, and *Porites*, as well as *Halimeda*, sponges, urchins and other invertebrates, at 20–25 metres depth. The asymmetry of the (former) lagoon seems due to active sedimentation in the shallow northern part, which lies on the leeward side, and is thus sheltered from currents generated by the southerly monsoon. Saya de Malha shows that a drowned atoll is not necessarily an extinct atoll: although its rim no longer attains the sea surface, active growth and sedimentation continue.

Drowned atolls are frequent in the Carolines (Tayama, 1952, p. 255), as at McLaughlin Bank and Gray Feather Bank. Velasco Reef, Palau, is another example (Tayama, 1952, Map 17), with a rim drowned under 10–15 metres of water, enclosing a lagoon 40 to 50 metres deep. Pulap Atoll in the Carolines (*ibid*., p. 163) is half submerged, the eastern side being entirely under water. The Geyser Bank, between the Comores and Madagascar (Fig. 127; see Fig. 60 for location), is a half-drowned atoll, and the nearby Zélée Bank is completely drowned.

On the Mid-Pacific Mountains, north of the Marshall Islands at 22°N, Darwin Guyot (Ladd *et al*., 1974), under 1266 metres of water, is very interesting because, in spite of its great depth and age, it has kept an atoll shape with a peripheral rim, and a central depression approximately 18 metres deeper. Shallow-water rudists of Cretaceous age have been dredged from the rim and the 'lagoon'. As it seems to be the oldest structure of this kind not to have been covered by later deposits, its name, coined by Ladd, Newman and Sohl, is certainly welcome.

Fig. 127 Part of Geyser Bank, western Indian Ocean between Mayotte and Madagascar, a tilted and partly submerged atoll belonging to the Comores chain (see Fig. 60) (Photo A. Guilcher)

4.10. Double- and triple-ring atolls

These frameworks resemble the double and multiple barrier reefs discussed in section 3.5 of this chapter, the difference being the absence of central high island. Nukuoro in the East Carolines (Tayama, 1952, p. 100) has an inner ring of lagoon pinnacles, and Greenwich, in the same group (Tayama, 1952, p. 155) has at least one and perhaps two inner rings, in addition to the usual atoll rim. Explanations suggested by Tayama (p. 255–256) resemble two of those proposed here for double barriers: that currents entering the lagoons through passages favour the growth of pinnacles along a course more or less parallel to the rim; or that there have been sudden tectonic movements during the subsidence. The first explanation if favoured by Kwajalein Atoll, Marshalls (Tayama, 1952, map 18 and Fig. 97 B, p. 111) and Ant Atoll, Caroline (Tayama, 1952, Fig. 97 A, p. 110), where swarms of pinnacles spread laterally from the inner ends of deep passages across the atoll rim.

4.11. Giant atolls

Atolls as large as Kwajalein, Rangiroa and Tijger (sizes at the beginning of this section 4) must have very large basements. It may be that some of them rest upon twin volcanoes instead of only one. Rangiroa for example has a size adequate for a basement made of volcanoes similar to the pair forming Tahiti Island (Tahiti Nui, Tahiti Iti *alias* Taiarapu). This idea receives support from the existence of twin atolls with the shape of an 8 at Woleai in the West Carolines (Tayama, 1952, Map 14). It is understandable that, in the course of subsidence, the central isthmus has widened gradually. Indeed, at Rangiroa there is a remnant of such an initial shape. On the other hand, Kuenen (1933, 1950) explained Tijger Atoll as 'topping a geanticline like a chimneypot standing on a roof, but it is much broader than the crown of the ridge and has evidently spread outward on its own screes'.

4.12. **Faros** (or faroes)

Although they have more or less similar counterparts in other areas, the faros off the west coast of India whence the name comes, are a unique assemblage which must be described first (Figs. 128 and 129). The Southern Maldives do not show quite typical features, but north of the Veimandu Channel they include:

(a) Two large circular groups of atolls stretching 260–300 kilometres north–south and about 90–100 kilometres east–west.

(b) Each of these two groups contains seven or eight large or very large atolls, 30 to 80 kilometres long, the lagoons of which being generally 35–70 metres deep. Between these atolls are channels generally more than 400 metres deep. Comparable depths occur in the centre of the two large circular groups.

(c) Faros occur on the edges of the large atolls. Some are 10 kilometres or more in length, but the majority are smaller. The smallest are almost circular, but others are elongated parallel to the edges of the large atolls. Their lagoons vary in depth betwen 4–6 metres and almost 40 metres. Three of the large atolls of type b contain many faros in their lagoons in addition to those occurring on their outer edge.

Agassiz (1903b) and Gardiner (1903) have accurately described the Maldives, but, as Davis (1928, p. 526–527) has said, their tentative explanations of the formation of these reefs are inadequate. Gardiner particularly tried to escape the necessity for subsidence, and he supposed an accumulation of deep-sea corals (?) over a pre-coralline basement, capped by shallow corals which would have created large atolls of type b. One interesting idea he suggested is, however, that the peripheral growth of these circular groups has prevented the development of reefs in the middle. More recently, Purdy in a brief abstract (1981) deduced, from 'seismic stratigraphy and results of a single well', that the Maldives developed during a long-continued subsidence of an Eocene volcanic substrate, leading to an accumulation of more than 2100 metres of shallow-water carbonates. He considered that 'the reef development took place in a decidedly non-Darwinian sense', since it was 'not accompanied by a genetic succession of fringing-barrier-atolls'. Purdy's con-

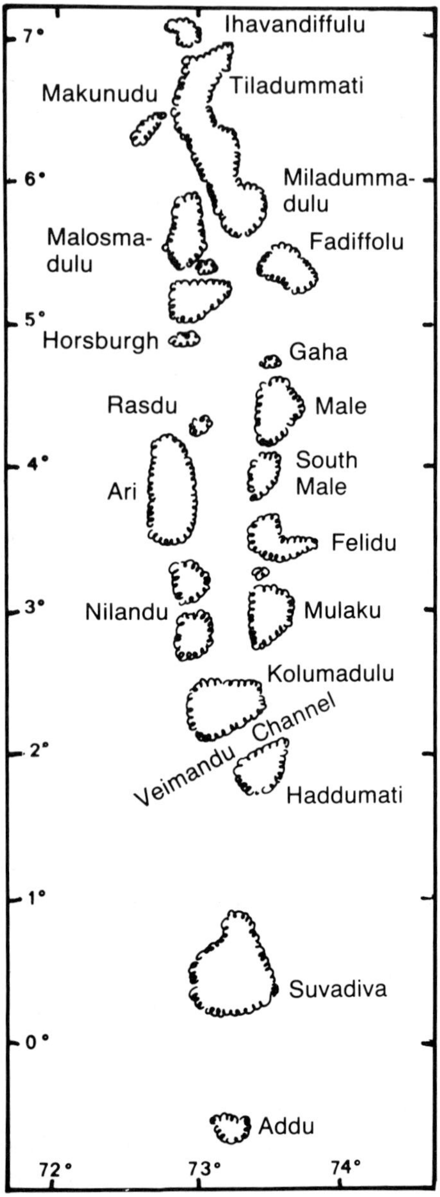

Fig. 128 General outline of the Maldives, eastern Indian Ocean. Only atolls are shown; for individual faroes at Tiladummati Atoll, see Fig. 129

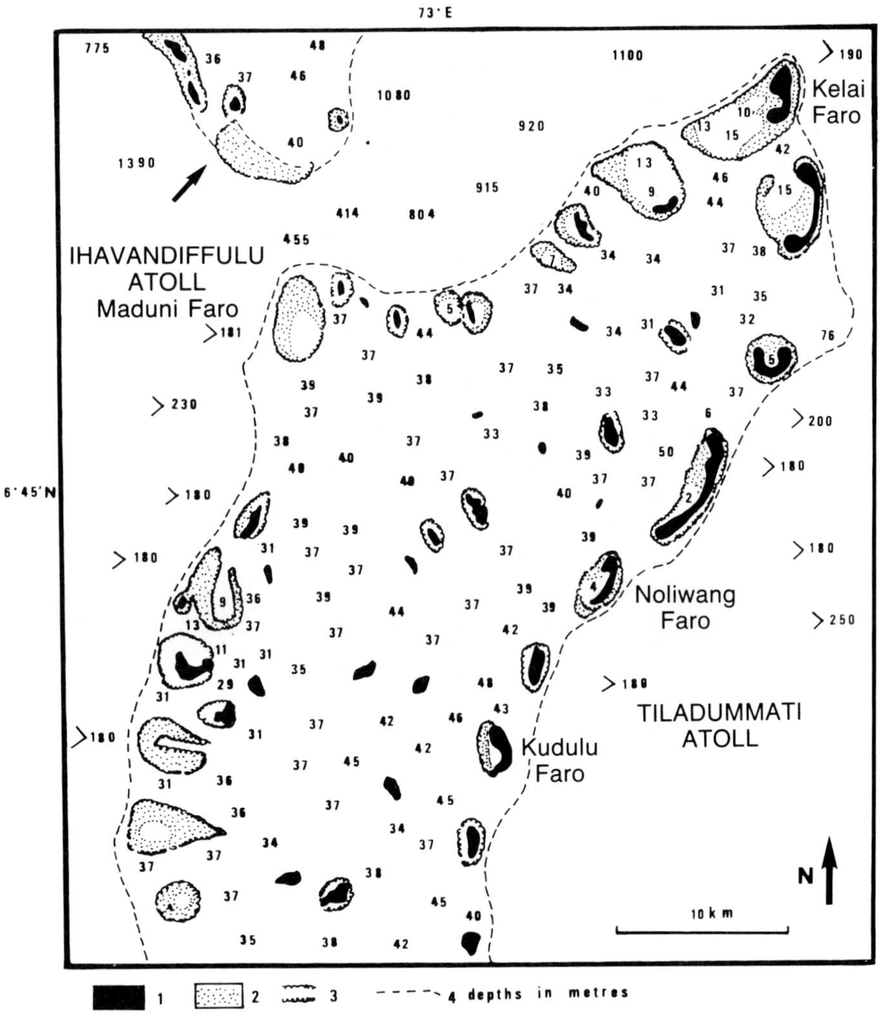

Fig. 129 Northern half of Tiladummati Atoll, Maldives, with faros. Data from Admiralty Chart 66 A. 1: islands; 2: very shallow; 3: edge of reefs; 4: limit of atolls

cept is not explained in sufficient detail to allow a sound opinion. Apart from the prolonged subsidence, which is essential, the problem of the formation of a and b types seems to require further information.

The small faros, which remain the most curious features of the Maldives, have horseshoe shapes, with the convex side facing the open sea, and thus seem (Guilcher, 1971, p. 80) to be related to the surf generated by the alternating monsoon winds on both sides of the b type atolls, while the faros situated in the middle of these atolls are almost symmetric. The same author (1963) has also drawn the attention to the similarity of the Maldive faros with

Fig. 130 Eastern barrier reef of New Caledonia, Melanesia. Doiman Reef off Tuo. Faro-like features related to wave refraction alongside a passage through the barrier (Photo A. Guilcher)

Fig. 131 Faros in the Great Southern Lagoon of New Caledonia (Photo A. Guilcher)

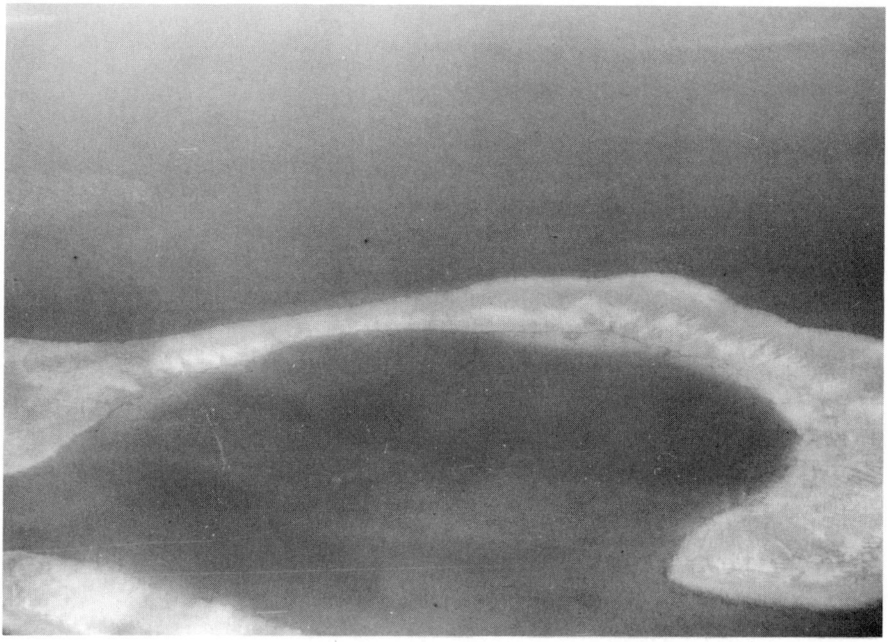

Fig. 132 A closer view of Tootira Faro, in the same lagoon as Fig. 131 (Photo A. Guilcher)

the horseshoe-shaped features found on the Mayotte and New Caledonia barrier reefs, and on the outer Great Barrier Reef (Figs. 88, 89, 90, and 130), which are obviously due to wave refraction on either side of passages. However, Stoddart (Guilcher, 1971, discussion p. 86) has not accepted this explanation: he insisted on the differences in the frequency of faros from north to south in the archipelago, probably resulting from contrasts in the rates of reef growth. So that the Maldives and their faros still remain open to discussion.

There are other features in the world which very much resemble the Maldivian faros, particularly on the Tagula barrier reef southeast of New Guinea (Daly, 1934, p. 222); around the large Suvadiva Atoll in Indonesia (Kuenen, 1947, p. 22, and 1950, p. 464); at Vanua Levu in Fiji (Kuenen, 1950, p. 470); and in the lagoon of Grand Récif Sud, New Caledonia, well inside the barrier reef and not parts of it (Guilcher, 1965a, plates I and III, and here, Figs. 131 and 132). The origin of most of these features remains controversial.

4.13. Elevated atolls

Several elevated atolls—or landforms considered to be emerged atolls—have been described from various parts of the Pacific: Maré, Lifou and Ouvea, the

Fig. 133 Makatea elevated atoll, between Tahiti and the Tuamotus. Four sets of notches can be seen in the cliff (Photo A. Guilcher)

three Loyalty Islands near New Caledonia and parts of this Territory; Makatea between Tahiti and the Tuamotus; Nauru between Tuvalu and the Solomons: Rennell in the Solomons; Kita Daito Zima and Minami Daito Zima off southern Japan (Doumenge, 1963; Chevalier, 1968; Leontyev and Medvedev, 1972; Kaplin *et al.*, 1973; Bourrouilh *et al.*, 1976; Gaven and Bourrouilh, 1981; Montaggioni, 1985; general review in Bourrouilh, 1977a). All these frameworks rise much higher above present sea-level than the slightly emerged atolls of the Tuamotus, Aldabra and Europa. Makatea rises 113 metres, Rennell 200 metres, Nauru 65 metres, Maré 138 metres, Lifou 60 metres, and the two Japanese islands approximately 50 metres; Ouvea is tilted, with a northwestern part still a living reef while the southeast has been uplifted 46 metres. In all these islands the greatest altitudes occur on an outer rim, which stands above a central depression, as in modern atolls. At Maré, two small volcanic cores are visible in the central depression, in other words the volcanic basement outcrops. The outer edges consist generally of steep cliffs, with sets of marine notches which bear marine deposits in several instances, e.g. at Makatea (Fig. 133). At Maré the peripheral rim includes smaller, shallower depressions which have been interpreted as fossil faros by Chevalier (1968), following Guilcher (1963). All the raised atolls have been strongly karstified, with caves sometimes tens of metres deep, and pinnacles that stand particularly high at Nauru (Fig. 134), with subvertical sides of the

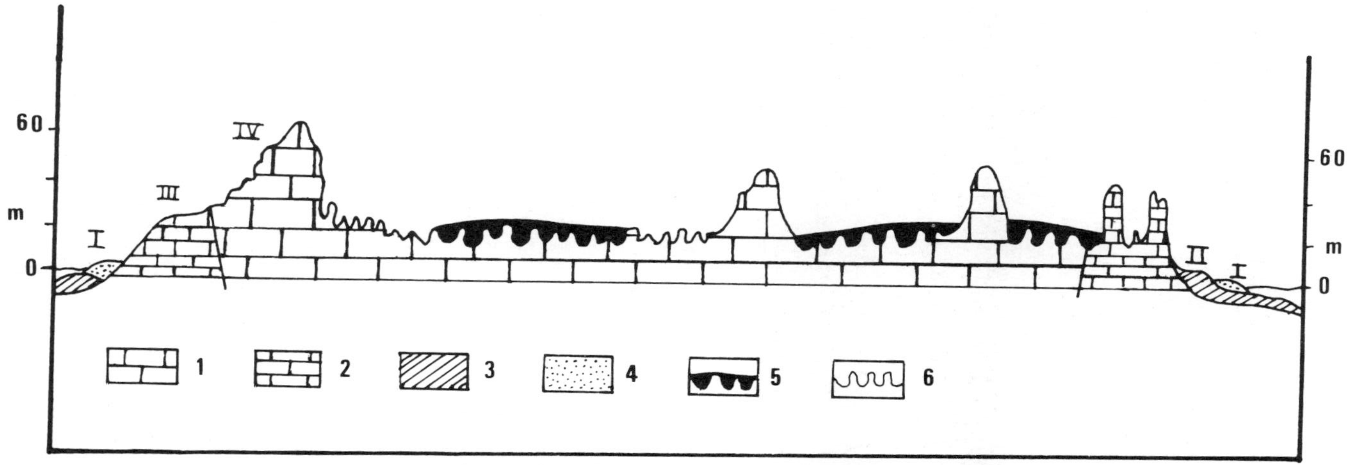

Fig. 134 Section across the elevated atoll of Nauru, Central Pacific, according to Leontyev and Medvedev (R V *Mendele'ev* expedition, 1971). 1: inner lagoon coral reef; 2: circular fossil rim coral reef (both assumed Paleogene, and dissected probably in the Miocene as tower karst); 3: Pleistocene reef; 4: loose Holocene deposits; 5: phosphate-filled karstic holes (lapiés); 6: karstic holes exposed by phosphate exploitation; I: Holocene terrace (3.5–4 metres); II, III, IV: Pleistocene notches at 10–11, 24–25, and 60 metres above sea-level

Fig. 135 Makatea elevated atoll. Exploitation of phosphates between the feos (karstified pinnacles). The phosphate deposits were exhausted in 1965 (Photo A. Guilcher, taken in 1963)

same type as in continental tower karsts (Turmkarsts) of China, Southeast Asia, Cuba, New Caledonia, etc. Dolomitization has occurred. In the depressions, the smaller karstic holes are being filled with thick phosphate deposits (now exploited) at Makatea (Fig. 135) and Nauru (Fig. 134). The age of the older parts of the structures is Tertiary: Palaeogene at Nauru, at least Middle Miocene at Rennell, and Early Miocene at Makatea; while the volcanic core of Maré is Upper Oligocene or Early Miocene. Maré and Makatea would have been faulted (Chevalier, 1968; Montaggioni, 1985; not accepted for Maré by Gaven and Bourrouilh, 1981). The peripheral notches are Pleistocene and perhaps older (precise dating at Makatea and Maré for the lower, Pleistocene ones). Fringing reefs grow now at sea-level around them.

As for atolls lying presently at sea-level, there are two interpretations: the Darwinian, and the karstic saucer theory. The first one was favoured by Chevalier (1968) at Maré, and the second one by Montaggioni at Makatea (1985; see here Fig. 136). Bourrouilh (1977) extends the karstic explanation to all these structures as well as to the atoll-like features in the Fiji Islands (Chapter three, section 5). Karstification is accepted by all, but opinions differ about the origin of the central depression. For phosphate formation, the theories proposed for Mataiva (section 4.8 above) have also been suggested.

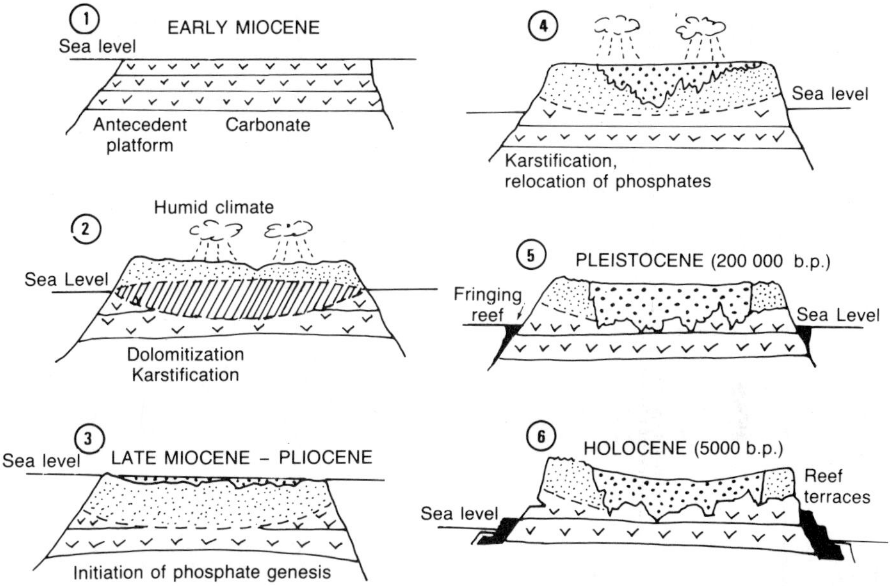

Fig. 136 Evolution of Makatea island as proposed by Montaggioni (1985), simplified. An example of the karstic saucer theory as an explanation of elevated 'atolls' (Bourrouilh, Purdy and others)

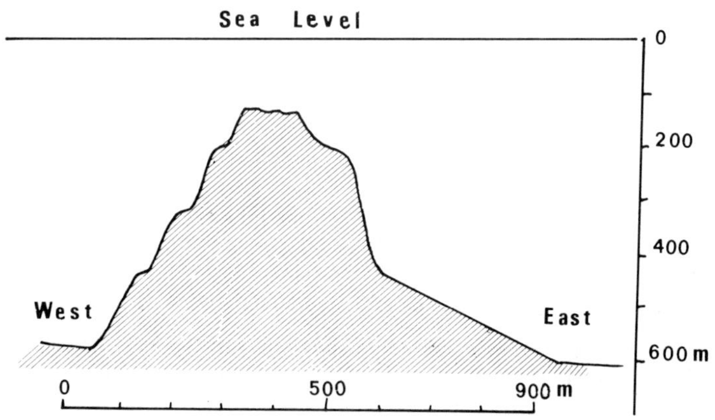

Fig. 137 Echo-sounding profile by R V *Calypso* across Shab Suleim ridge, Red Sea off Saudi Arabia, 1 kilometre NNW of last emergent reef shown on Fig. 138. Lengths and depths at same scale. After Guilcher (1955)

5. RIDGE REEFS

Ridge reefs have not been recognized so far as a special type of coral reefs; but, following an opinion expressed in 1965 by Ph. Kuenen after he read Guilcher's report (1955) on the Farsan Bank in the Red Sea, it seems that they should be so.

It must be remembered first that the Red Sea, in which this type of reefs is found, is an ocean in the making, since it lies between the gradually diverging

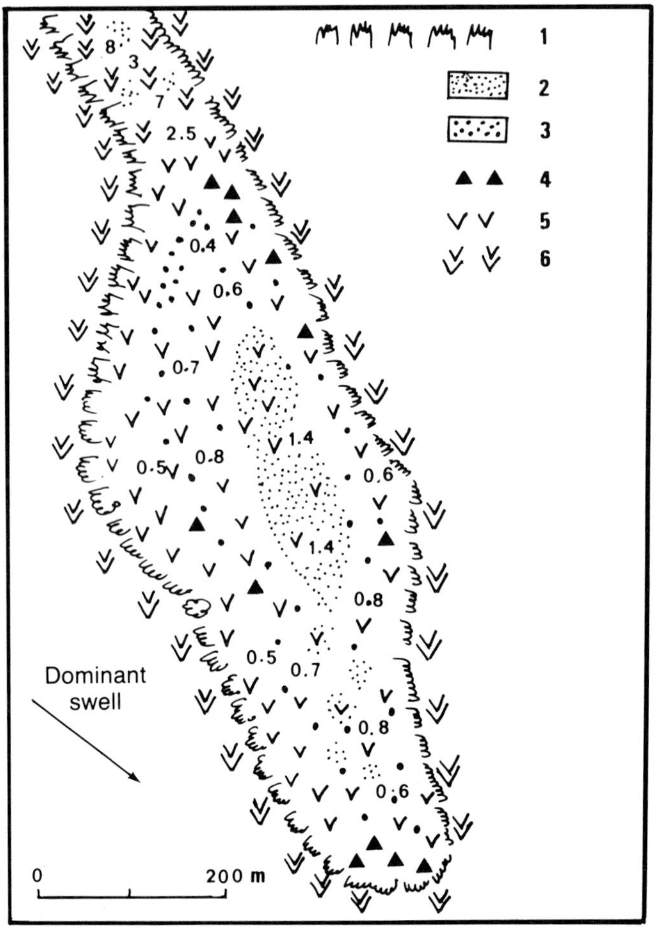

Fig. 138 Top of Shab Suleim ridge reef, Farsan Bank, Red Sea. Depths in metres. Extremely small tidal range. After Guilcher (1955) (R V *Calpyso* expedition, 1952). 1: reef edge (with spurs and grooves on western, exposed side; 2: sand on dead reef; 3: small detrital coral blocks; 4: big blocks; 5: living colonies; 6: continuous coral cover. Precipitous submarine slopes on both sides, descending to more than 500 metres, see Fig. 137

African and Arabian plates, and, in the north, the Sinai platelet. Recent studies (e.g. Pautot, 1983; Bonatti, 1985; Pautot *et al*., 1986) have shown that its evolution consists of progressive oceanization, proceeding from south to north, along a central graben originally surveyed by R V *Calypso* (Nesteroff, 1952), through a series of 'cells', located over more or less regularly spaced 'hot points'. The general trend of these cells and of the graben is northwest to southeast, a direction which is also followed by the ridge reefs found on either side of the graben.

Farsan Bank, on which the biggest reefs of this type are situated, is a huge area off the Asir coast of western Saudi Arabia, over 580 kilometres long and 100 kilometres wide, between the Tihama (coastal plain) and the central graben. It consists of alternations of hollows several hundreds of metres deep and coral reefs rising abruptly to the surface. In the area investigated off Lith (Guilcher, 1955; Nesteroff, 1955), the reefs consist of long ridges (e.g. Shab Suleim, Shab Sahabak) running NW–SE, i.e. in the general direction of the Red Sea, the depth on either side of Shab Suleim being between 600 and 700 metres, beside extremely steep slopes (Figs. 137 and 138). Similar ridges have been described off Jeddah by Mergner (1967) and called by him outer reefs (Aussenriffe in German). These features have been interpreted by Guilcher (Fig. 139) as coral structures which originated on small horsts, and then grew upward during the progressive subsidence of Farsan Bank. The present state of research does not allow to know what happened during the Quaternary

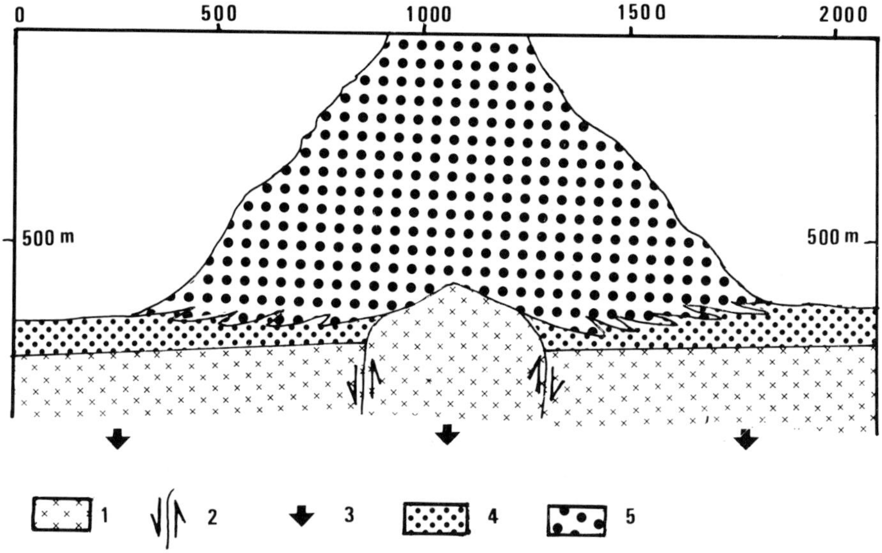

Fig. 139 Supposed formation of the Farsan Bank ridge reefs (Shab Suleim etc). From Guilcher (1955) (RV *Calypso* expedition, 1952). 1: basement; 2: horst formation; 3: general subsidence; 4: bottom sedimentation, initially continental, later marine; 5: corals and associated organisms. Depths and lengths at same scale

Fig. 140 Abulat Island (emerged corals on Farsan Bank), broken into horsts and grabens by faulting in a NW–SE direction (Photo A. Guilcher)

oscillations of sea-level, but the evolution suggested is not incompatible with these oscillations.

This tectonic interpretation of the ridge reefs of Farsan bank is not only in agreement with the general evolution of the Red Sea; it also fits the morphology and structure of Abulat Island (Guilcher, 1955), which lies closer to the coast, on a higher step than Farsan Bank, and consists of emerged corals, heavily faulted, again in a NW–SE direction (Fig. 140). Abulat is thus a raised, emerged ridge reef.

In addition, at the northwestern end of the Red Sea, the southern tip of the Sinai Peninsula, which is situated at the triple junction between the African and Arabian plates and the Sinai platelet, consists of emerged coral reefs that have been cut into small horsts and grabens, which run in the same direction (Gvirtzman and Buchbinder, 1978; Guilcher, 1979a) (Fig. 141). These frameworks continue into the shallow area at the entrance of the Gulf of Suez, where some of them have been interpreted as the outcome of tidal currents running to and fro (Roberts and Murray, 1983). From a discussion on that point (Guilcher, 1986, comments and reply), it results that the ridges are structurally controlled, being thus tectonic ridges, of smaller size than those on the Farsan Bank; the submerged ones may well be influenced in their growth by local tidal currents, which do not exist elsewhere in the Red Sea, where the tidal range is extremely small, particularly over Farsan Bank.

Fig. 141 Emerged fossil ridge reef at Ras Muhammed, the southernmost tip of the Sinai Peninsula. Scale given by vehicles in background (Photo A. Guilcher)

Ridge reefs also occur in shallow water on the western side of the Red Sea, along the coast of Sudan, where El Rabaa (in Guilcher, 1986) has interpreted them as resulting from slow subsidence of horsts; and this is in agreement with Braithwaite (1982). It is even more interesting that, like Farsan Bank off Saudi Arabia, the Sudanese ridge reefs extend to the deep sea in front, as shown by Schuhmacher and Mergner's (1985) excellent study of Sanganeb Atoll. This atoll (Fig. 142) is elongated, and extends in a north–south direction, parallel to the coastal ridge reefs; it rises from depths of 550 metres in the east and 750–825 metres in the west. It is interpreted by the German authors as having been built on a horst: the thickness of the coral reefs is a matter for speculation, but the tectonic interpretation is in total agreement with the structural Red Sea context. Sanganeb Atoll, although it resembles other atolls in the world in its shape and depth of lagoon, thus belongs to a quite special subtype: a ridge atoll. It is the only atoll investigated in the Red Sea. Another reef, called Green Reef, southeast of Suakin, Sudan, resembles an atoll and was visited by RV *Calypso* in 1952 (Guilcher, 1955, p. 89–90), but this has not yet been studied in detail.

Summarizing these results, ridge reefs are certainly widespread in the Red Sea, and include as subtypes the shallow and emerged ridge reefs on the

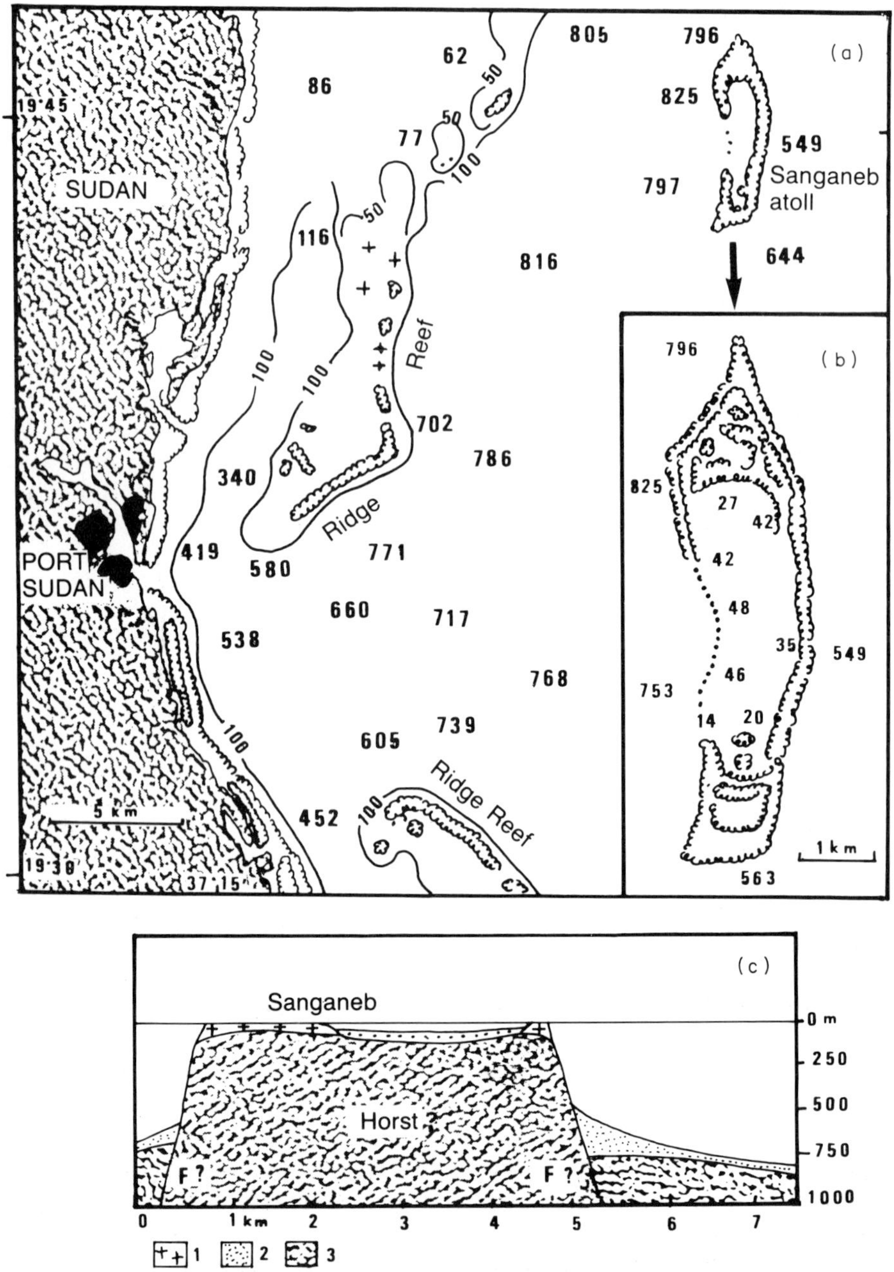

Fig. 142 Coral reefs off Port Sudan, western Red Sea. Mainly after Schuhmacher and Mergner (1985); ridge reef interpretation by Guilcher. (a) General map. Depths and contours in metres (b) Sanganeb Atoll, enlarged from (a). (c) section across Sanganeb Atoll, structure as interpreted by Schuhmacher and Mergner. 1: built-up reef; 2: other coral sediments; 3: rocks of unknown nature; F: faults. The embayments at and north of Port Sudan are typical marsas or sharms (see Figs. 37 and 38)

eastern, western and northern sides, the huge ridge reefs in deeper water off Arabia and Sudan, and one ridge atoll off Sudan.

As many other features concerning coral reefs, the Red Sea ridge reefs were mentioned by Darwin (1842, p. 192), on the basis of information from Captain Moresby. He wrote: 'some of them are very long in proportion to their width . . . many of them rise to the surface . . . they extend parallel to the shore . . . the sea is generally profoundly deep close to them, as it is near most parts of the coast of the main land . . .'. These data were unfortunately forgotten for a very long time after him. So that Darwin deserves the name of ridge reefs' grandfather.

Are ridge reefs represented in other areas? Braithwaite (1982), referring to Heidecker (1973), suggested that 'there are analogies between the tectonic setting of the reefs here (near Port Sudan) and those off the Queensland coast'. But, although Heidecker gave evidence for tectonic and structural control of recent reefs off northeast Queensland, it does not seem that these features really resemble the Red Sea ridge reefs. So far as is known, this type of reefs still lacks counterparts elsewhere.

6. SPECIAL REEFS

There are other reefs, which may be very large, and do not fall into the preceding categories.

6.1 Bahaman reefs

Although the Bahaman Banks lie in an environment where temperature, salinity and low turbidity are favourable for reef growth, they are not true coral reefs, in spite of the fact that corals grow on their upper edges. They will be included in this general review for comparison, with a selection from the large bibliography on the Bahamas (Illing, 1954; Newell and Rigby, 1957; Newell *et al*., 1959; Hoffmeister *et al*., 1967; Milliman, 1967; Bourrouilh, 1974, 1977b 1982; Mullins and Lynts, 1977; Mullins *et al*., 1982; Hine *et al*., 1981; Garret and Gould, 1984; Freeman-Lynde and Ryan, 1985).

These banks extend over 10 degrees of longitude off Florida, Cuba and Hispaniola (Fig. 143). The size of the Great Bahama Bank (550 × 330 kilometres), and the Little Bahama Bank (250 × 70 kilometres) exceeds the largest atolls, including those in the Maldives and the Great Chagos Bank; they decrease in size in the southeastern banks, where the biggest one, Caicos Bank, is still 100 kilometres long. Like atolls they bear low islands on their periphery, and their inner parts are submerged; but these inner parts are generally much shallower than in atolls, reaching depths of only a few metres, so that they are submarine platforms rather than lagoons. These inner platforms, particularly on the Great Bahama Bank (Fig. 144), bear oolites of cryptocrystalline aragonite, a sediment which on the western parts of the bank forms shallow, extremely mobile ridges, in which the asymmetry of the ripples

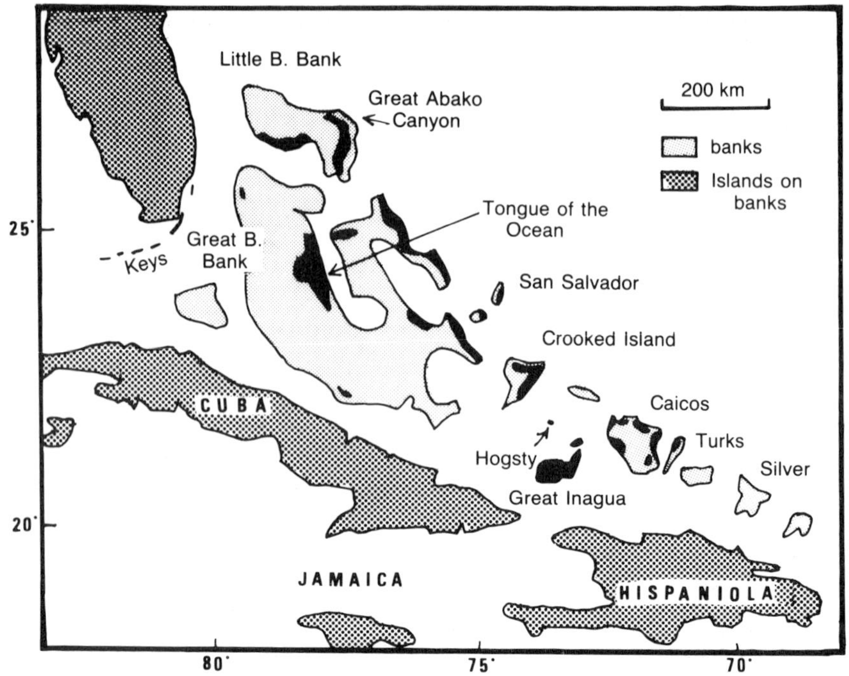

Fig. 143 General setting of the Bahaman Banks

changes with the tide. The other sediments on the same bank are more stable, and include mixed skeletal sands, pellets, 'grapestones' (aggregates), and calcareous ooze or mud in the deeper parts. True coral reefs are generally restricted to the top of the eastern, windward slope of the bank.

The islands, always located on the sides, never in the middle, lie generally on the eastern edge. Although usually low, they sometimes bear Pleistocene aeolianites, e.g. at New Providence Island, where these consolidated dunes reach 25 to 30 metres at Nassau City, standing above and behind Pleistocene beach ridges. The largest island, Andros, displays an extremely complicated pattern of low Pleistocene karstified limestones (including 'blue holes', see Chapter three, section 2), with multiple long, sinuous consolidated ridges lying just above present sea-level, which are fossil marine ripples. Hurricanes now create lenticular mud bodies called cyclonic trails, which are superimposed over the fossil features.

In spite of their differences with 'ordinary' coral reefs, the Bahaman banks have a few counterparts. In the shallow Gulf of Batabano, 7 to 12 metres deep, between mainland Cuba and the Isle of Pines, the distribution of sediment types is, according to Daetwyler and Kidwell (1959), much as on the Great Bahaman Bank (compare Fig. 144): in the north, calcareous muds occur; the central part is occupied by oolites and grapestones; and in the

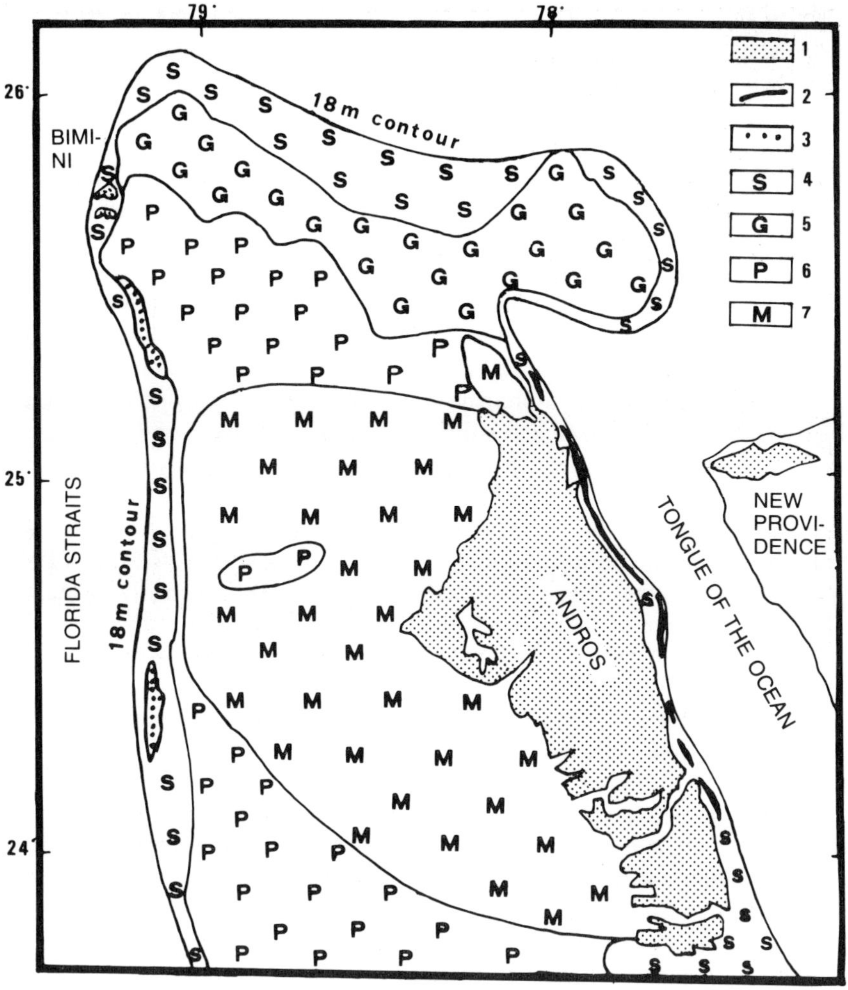

Fig. 144 Sediments on the northwestern part of the Great Bahaman Bank. After Newell *et al.* (1959). 1: islands (Andros contours are simplified); 2: coral reefs (mainly *Acropora palmata*); 3: unstable oolite; 4: mixed skeletal sand; 5: grapestone sand; 6: pellet sand; 7: muddy sand and mud

south, skeletal calcareous sediments and a row of coral reefs are found on the edge of the very steep slope descending to the depths of the Caribbean Sea. On the other hand, the precipitation of oolites, although infrequent on true coral reefs, is not restricted to the Bahaman Banks and the Gulf of Batabano, since it occurs in the shallow waters of the southern part of the Persian or Arabian Gulf, and also on the western Florida shelf and the northern shelf of Yucatan (Van Andel and Veevers, 1967, giving references).

A framework resembling an atoll more than the other Bahaman features is Hogsty Reef, in the southeastern part of the group, described by Milliman (1967). It has the shape of a small (9 × 5 kilometres) atoll, with a shallow (6 to 8 metres) lagoon, open on its leeward side. But its rim appears to be a Pleistocene dune, and hermatypic organisms, which now flourish on it, seem to have been only recently (re?)introduced. Hogsty Reef is thus an atoll in terms of configuration, but quite different in structure from the usual type.

The submarine outer slopes of the Bahaman Banks are extremely steep, and drop precipitously (commonly 28 to 40°) to great depths (more than 3500 metres), as around atolls. But the most curious aspect of the Bahamas is the existence of the Tongue of the Ocean, a deep-water channel, wider than a submarine canyon, which merges downwards into the Providence Channel with up to 4800 metres of submarine relief. On the eastern side of the Little Bahama Bank, the Great Abaco submarine canyon is not so wide, but is also more than 3600 metres deep in its outer section. Such canyons are never found around true atolls.

Before the origin of the Bahaman Banks is examined, their similarity with the nearby peninsula of Florida must be clarified. As Hoffmeister *et al*. (1967) have pointed out, the Pleistocene sediments in Florida form, from east to west, a mirror image of their Recent Bahaman counterparts of the other side of the Straits of Florida: the Pleistocene Floridan coral keys, a Pleistocene oolitic ridge which corresponds with the present unstable Bahaman oolites, then a Pleistocene bryozoan facies equivalent to the facies existing in the middle of the Great Bahaman Bank.

Several deep drillings have been made in different parts of the Bahaman Banks, and there is no doubt that they are much thicker than the thickest investigated atoll, Enewetak. The four boreholes in the northwestern part of the Great Bahaman Bank (Mullins and Lynts, 1977) show that the Cenozoic and Cretaceous sediments are at least 4800 metres thick; the pre-Jurassic basement being at −6000 or −7000 metres. These figures require a long-continued subsidence, which was initiated at a time when the Atlantic Ocean was just opening between America and Africa. Deep channels such as the Tongue of the Ocean are certainly very old, but whether they are as old as the opening of the Atlantic remains a matter of discussion. Several models have been proposed, including one by Mullins and Lynts (1977) (Fig. 145), according to which the depressions were initiated as grabens when the platform was still part of Africa; the platform would have later drifted and rotated during the Cretaceous and Early Tertiary, with subsidence testified by the thick sedimentation on the banks. Fig. 145 shows the connection with Florida. In the Great Abaco canyon (Mullins *et al*., 1982), the steep walls consist of Cretaceous massive limestones, and the orientation of the canyon is structurally controlled. The absence of filling by sedimentation in that canyon may be due to episodic erosion along the axis during the Pliocene and the Pleistocene, but erosional activity is presently small. In the Florida Straits, on the western edge of which coral growth now plays a larger part than in the bank

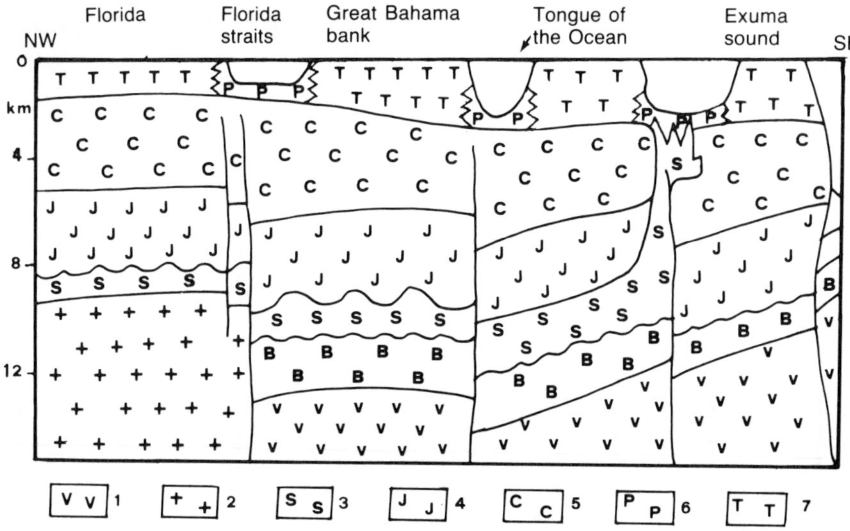

Fig. 145 The Bahaman Banks, northwestern half, and Florida. Interpretative cross-section. After Mullins and Lynts (1977). 1: oceanic crust. 2: Pre-Triassic igneous and metamorphic rocks. 3: Jurassic salt. 4: Jurassic carbonate deposits. 5: Cretaceous limestones. 6: Tertiary pelagic limestones. 7: Tertiary shallow water limestones. Depths in kilometres

environment, the absence of filling is more understandable than in the Tongue of the Ocean and other similar troughs, because of scour by the Gulf Stream.

Leg 101 of the *Joides Resolution* Ocean Drilling Program (ODP) in 1985 sought to collect firm data on the Bahaman structure and its evolution, but technical problems hampered drilling in the Florida Straits and other troughs, so that the interpretations are still hypothetical.

6.2. Houtman's Abrolhos

These coral islands (Fig. 146), the southernmost ones in the Indian Ocean, are not only unique in their location off a western continental coast (apart from some poor reefs off the west coast of Africa); they are also unusual in their shape, as they are neither atolls nor banks nor fringing reefs. Darwin (1842) left their classification in doubt, and Dàvis (1928, p. 204) hesitated between 'bank reefs' and 'bank atolls'. More recently they have been described accurately by Teichert (1947) and Fairbridge (1946–1947), and their ecology defined by Hatcher (1985). The reason for their good development at 28°30′S–29°S is the warm Leeuwin current (Chapter one, section 2).

Houtman's Abrolhos, situated some 70 kilometres offshore, rise near the outer edge of the Australian shelf from depths of 50–55 metres. They comprise four groups of islets: the two northernmost are somewhat shapeless,

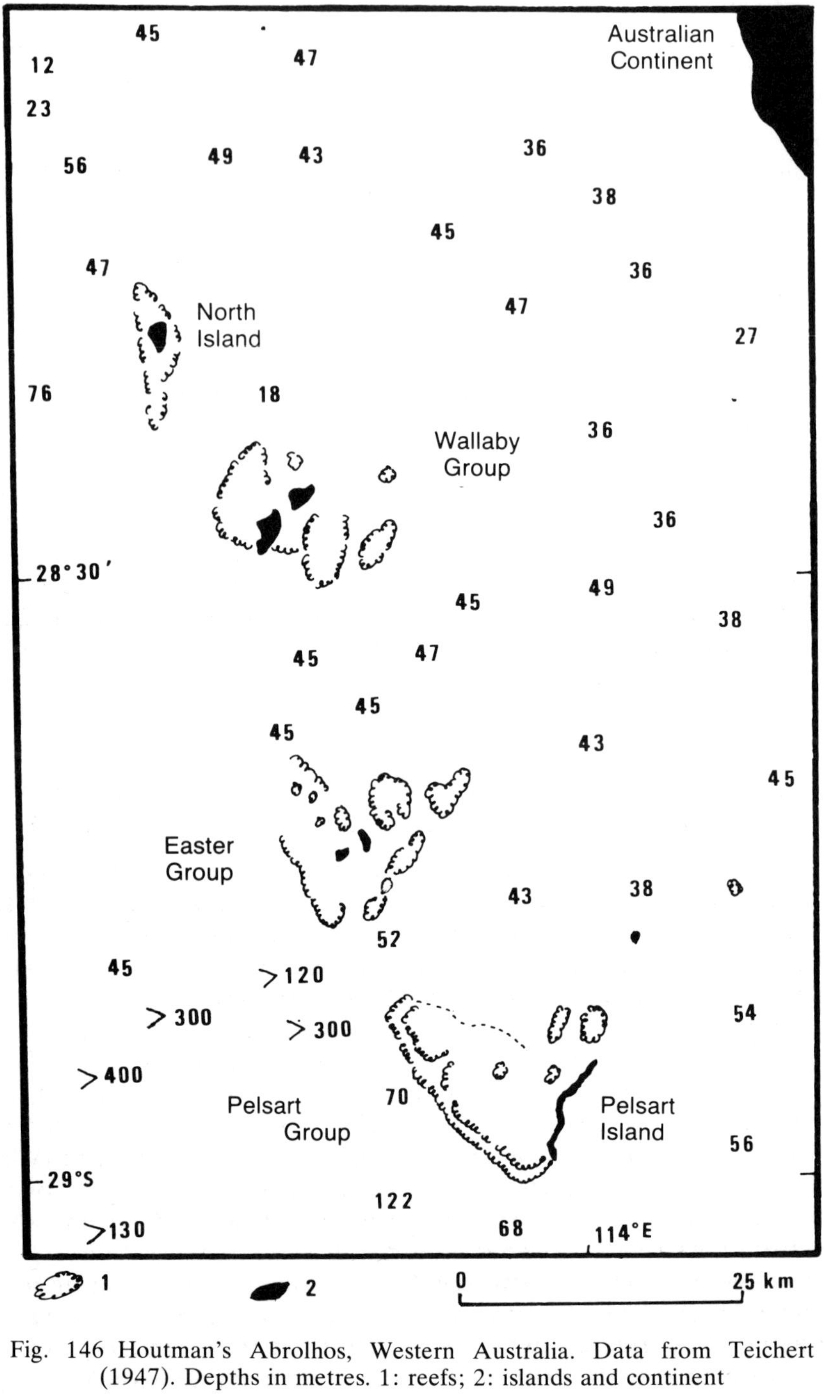

Fig. 146 Houtman's Abrolhos, Western Australia. Data from Teichert (1947). Depths in metres. 1: reefs; 2: islands and continent

while Easter Group and Pelsart Group, in the south, are V shaped, with the apex in the south, and a wide opening with many coral patches or knolls in the north. The structure is complex. The foundation in each case is a Pleistocene coral reef limestone, rising up to 1 or 2 metres above high tide level in its highest parts. It is covered in many places by a shelly limestone, above which a shingle limestone is found, and finally a dune limestone reaching 10 metres in thickness at West Wallaby Island. Subaqueous shingle deposits, beach ridges of coral shingle, small dunes in the Wallaby Islands, and guano, represent the modern sediments. The growth of corals is luxuriant in many places, and the algal flora is extremely diversified, with both tropical and temperate components. From a geomorphological point of view, the most interesting group is certainly Pelsart, in which the eastern branch of the V forms an extremely narrow (30 to 600 metres wide) emerged island 11 kilometres long, in which various sedimentary components outcrop, whereas the western branch is an outer reef at low water level with a few remnants of a higher, former reef.

It is surprising to find such a complicated reef on the outskirts of the coral world. Generally the reefs in such a situation are simply fringing reefs, as at Lord Howe in the Tasman Sea (Chapter four, section 1.3); but Bermuda reefs, also at the limits of coral reef growth in the North Atlantic, are equally somewhat complex, being classified as a bank-atoll by Davis (1928, p. 212), although this is questionable. The clue to this problem may lie in the excellent hydrological environment in both cases, with warm currents favouring coral growth in spite of the latitude.

7. COMPARISON BETWEEN THE INDO-PACIFIC AND ATLANTIC CORAL WORLDS, AND PROBLEMS OF THE BRAZILIAN REEFS

The main contrast between the Indo-Pacific and the Atlantic coral worlds lies in the number of hermatypic coral genera in the two areas (Fig. 2). The heart of the coral kingdom is situated in Melanesia, Indonesia and the Eastern Indian Ocean. From that area the number of genera decreases outwards, especially in Polynesia, and, as a whole, the Western Atlantic is much poorer.

However, the Caribbean reefs are by no means pale images of their Indo-Pacific brothers (Milliman, 1973). On their outer, windward sides, they often display vigorous growth of *Acropora palmata* (Fig. 8A), with *A. cervicornis* a little lower, which can match similar parts of reefs in the Indo-Pacific. But, as Adey and Burke (1976) and Laborel and Bouchon (1986) have pointed out, such patterns are replaced in other sites by fleshy algal pavements, also associated with strong wave energy, as in Martinique where they seem to have recently replaced *A. palmata*. Invasion of soft algae, which change the nature of the reefs, may be related to an increase in turbidity, resulting, in several cases at least, from modern erosion on the nearby subaerial slopes (see Chapter five). Yet in the heart of the coral kingdom, the reefs of southeast Asia have to withstand heavy sedimentation from large rivers, so that the contrast in this respect is far from being complete. On the other hand, many Caribbean reefs are subject to more severe hurricanes than a large majority of Indo-Pacific reefs.

Although present in the Caribbean, there are very few atolls in comparison with the Indo-Pacific (17, or even 10, instead of more than 300), and often not so typical as they are open on their leeward sides (Fig. 104); this shape is also represented in the Pacific, e.g. at Tarawa (Fig. 103), but is rather rare there. A reason for the scarcity of atolls in the Caribbean may be that migration of a plate over hot spots, which appears to be the origin of atoll chains in the Pacific (Fig. 59) and possibly also in the Indian Ocean (Fig. 60), has not been effective in the Caribbean. The movement of the Caribbean plate between the Nazca, Cocos and American plates has not resulted in atoll chains, apparently because of a lack of melting anomalies in the asthenosphere. Milliman (1977) remarked that, instead of subsidence, much of the Caribbean area had uplift and active volcanism during the Tertiary; but the same author and Supko (1968) expressed the counter opinion that atolls such as Courtown and Albuquerque, off the Nicaraguan shelf, were probably built on a subsiding basement as in the Pacific examples. Information on the deep structure of that area (DSDP, 1973; Matthews and Holcombe, 1985) is not conclusive in this respect. The Venezuela basin rests upon an oceanic-type crust, at least as old as Late Cretaceous, with a thick pile of Tertiary sediments above it—a structure apparently favourable to atoll genesis; but no typical atoll has been reported from that area. Conversely, the Columbia basin contains atolls off Nicaragua, but, in the Deep Sea Drilling Project, Leg 15 failed to obtain useful structural data in that basin: there was only one drilling, n° 154, which penetrated into thick volcanic sands, silts and clays, and stopped at 270 metres in them.

It may be that the karstic saucer theory (Chapter three, section 5) finds more support in the Caribbean, particularly in Belize barrier reef (Fig. 49A) than generally in the Pacific; although the evidence from Fiji (Ladd and Hoffmeister, 1945) appeared convincing (Fig.49B and C).

Outside the Caribbean, coral reefs are represented in the Atlantic along northeastern Brazil, where they have been described in detail by Laborel (1969; additional remarks in Guilcher, 1983, for the João Pessoa area). The development of the Brazilian reefs has been hindered by two obstacles (Fig. 147): in the north the Amazon River, at the mouth of which the coral growth is prevented by the freshwater and the suspended sediments supplied by the largest river in the world, so that reefs are absent north from São Luis de Maranhão; in the south, a double thermal obstacle: an upwelling in Cabo Frio area, and cold water deriving from the Falkland current. The southernmost boundary of the reefs thus lies between Rio de Janeiro and Santos. In the intermediate area, there is an impoverishment of coral in both directions: towards the north between Cabo São Roque and Maranhão, and towards the south from Cabo São Thomé near Rio. In addition, coral growth is still restricted by the sediment and freshwater supplied by Rio São Francisco and Rio Doce: so that, finally, the favourable areas are limited to the

Fig. 147 Coral reefs and their environment in Brazil. After Laborel (1969). 1: warm current. 2: cold current (Falklands). 3: heavy sedimentation from rivers. 4: main reefs

Natal–Recife coast on one hand, and the São Salvador–Abrolhos region on the other.

In the Abrolhos region (not to be confused with Houtman's Abrolhos in Western Australia), which includes the Abrolhos archipelago proper and the nearby Paredes reefs (Fig. 148), coral growth is rich. The reefs are everywhere at least 20 metres thick and include a large number of flourishing pinnacles (the so-called chapeiroes) which may coalesce laterally. Their surface is covered by thriving colonies of coral, with lateral overhangs, as in the best reefs of the Indo-Pacific region.

In other Brazilian areas, a common peculiarity of the reefs is their association with *arrecifes*, which are subparallel to the coastline and partly submerged lines of Holocene beach rock, dated at between 7000 and 2000 BP (Flexor and Martin, 1979). This type of reef occurs (Ottmann, 1960) mostly in the Natal–Recife area, where corals form fringing reefs attached on the

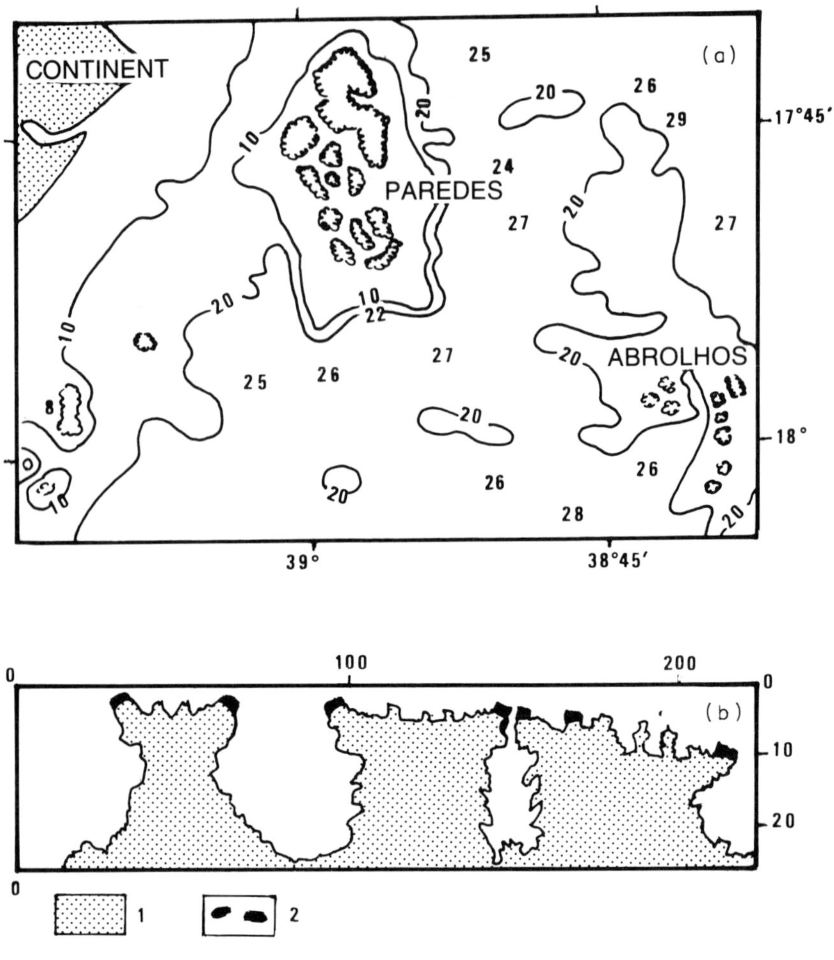

Fig. 148 The Abrolhos and Paredes Reefs, Brazil, after Laborel (1969). (a) general setting. Depths in metres. (b) section across a series of 'chapeiroes' in the Abrolhos, in process of coalescence. 1: built-up reef; 2: growing colonies

outer side of the seaward arrecife. Particularly near João Pessoa in Paraiba (Guilcher, 1983), the hermatypic corals are only scattered over the arrecife, whereas algae (e.g. *Caulerpa*, *Halimeda*) are predominant, so that the name of algal reef would be appropriate. The calcareous (95–98 per cent of carbonates) sediment consists essentially of *Halimeda* particles in many places. The turbidity of the water is high during the rainy season, which lasts for seven or eight months; mud thus carried to the reef gives a grey colour to dead *Halimeda* fragments, which strongly differ in colour from similar particles encountered on Pacific and Indian Ocean atolls. This turbidity is probably one of the causes of the poor development of hermatypic reefs.

The coral reefs existing along the Brazilian coast never go beyond the stage of fringing reefs, although they had ancestors during Pleistocene low sea levels (Laborel, 1969). Laborel also notes that, on islets lying far off the coast such as Fernando do Noronha and Rocas, there are not true coral reefs, but only scattered coral colonies which have not been able to build real reef frameworks. In spite of the name 'Atol das Rocas', and the circular shape of this structure with a very shallow central depression, less than one metre deep, it is not a corallian atoll since it is almost exclusively made of calcareous algae (Ottmann, 1962). In these sites, water turbidity cannot be cause of limited coral growth, so that the poor development of reefs in Brazil remains somewhat unsolved.

CHAPTER FIVE

Reef and Man

1. REEF DEGRADATION

In recent years, coastal geomorphologists have been increasingly interested in problems of coastal management, because coasts have been more and more subject to damage by man's activities, and these changes have been investigated by geomorphologists. In a similar way, the vigour and even the existence of coral reefs have been threatened by man, so that in the International Congresses at Manila (1981) and Tahiti (1985) such topics were widely discussed. A book dealing with coral reef geomorphology must therefore include a review of these various modifications, and of the efforts made to reduce or halt them, since without such management there will be no more reef geomorphology to study.

In a few areas, scientists considered that the situation is not dangerous. Such is the case at Madagascar, where Rabesandratana (1985) expressed the view that reefs are useful because they attract a large number of tourists, provide lime from broken shells, are a source of food with their fishes, gastropods, crustaceans and holothurians, and thus play a role in the socioeconomic activities of the country. No mention is made of any excess in exploitation nor danger of disappearance. Similarly, Mallik (1985) considered that Kavaratti and Kalpeni Atolls, Lakshadweep (formerly Laquedives, Indian Ocean) constitute 'an enormous potentiality . . . for industrial use (in) the manufacture of cement, calcium carbide, etc. . . . It is expected that exploitation of 1 or 2 metres of sand from the lagoon floor for industrial purpose will not affect the equilibrium, and attempts should be made to achieve an optimum depth of dredging'. In the South Pacific, the coral reefs of the

Solomon Islands 'are still close to their natural state except perhaps in localized areas' (Dahl, 1985), and in Vanuatu 'there has probably been little human impact' (Dahl, 1985).

On a world scale, the general opinion is, however, that reefs and their environment have deteriorated.

A first cause of destruction has been extensive illegal dynamiting as a fishing technique since the Second World War. This has affected many South Pacific Islands that were occupied by the US Army and Navy during the war, since, when they departed in 1945, American troops left behind them considerable amounts of explosives. They were particularly used by the Wallisians, who have, in the following years, completely devastated the lagoon between their island and its barrier reef. Some parts of the reefs of New Caledonia, Fiji, and Samoa have also been damaged in the same way (Doumenge, 1966).

Even outside the South Pacific, dynamiting has been reported from a wide range of reefs, such as those in Mauritius (Fagonee, 1985), Tanzania (Kudoja, 1985), Indonesia (Soegiarto, 1985), and Sri Lanka (De Silva, 1985). In several instances, overexploitation of living resources, without much dynamiting, is sufficient to create a hazard. Such is the case in the Maldives (Munch-Petersen, 1985), where traditional fishing, which exploited moderately reef fishes, turtles, molluscs, crustaceans, and octopi, has recently improved its catches considerably by the use of mechanized vessels and expansion through the archipelago to face the demand of local tourism and world market, taking a much wider range of marine products than the former Maldivian output. On the Chinese island of Hainan (Wu, 1985), where extensive coral reefs were reported in the 1950s, they have now largely disappeared or been replaced by algal covers. The damage is due mainly to the use of fine meshed nets which have 'removed all the fish', thus destroying the coral community.

Yet the impact on Pacific reefs and shores of the war itself has not proved, as a whole, very durable. The Japanese fleet sunk in the depths of Truk lagoon in the Carolines has been buried under a growth of marine organisms; some remnants of American lorries persist on the shores of Tarawa and other military landing sites; silhouettes of wrecked Japanese ships variegate the marine skylines of Guadalcanal and New Britain; and the American quays built on fringing reefs on the inner reaches of the Bora Bora lagoon have been quietly invaded by the nearshore vegetation.

Another potential threat is the possibility of establishing military bases in atolls. Such has been the case on Aldabra, Indian Ocean, a slightly emerged atoll of Sangamon age (Fig. 108), known to be the only undisturbed habitat of the Giant Land Tortoise in this ocean. Its bird population and flora are also most interesting. It was considered, in the middle 1960s, as a possible site for the development of a British military airfield, which would certainly have completely modified the environment (Stoddart, 1968).

A cause of coral reef destruction is the mining of coral and dredging of sand and gravel for construction and industrial use. In French Polynesia (Gabrie *et*

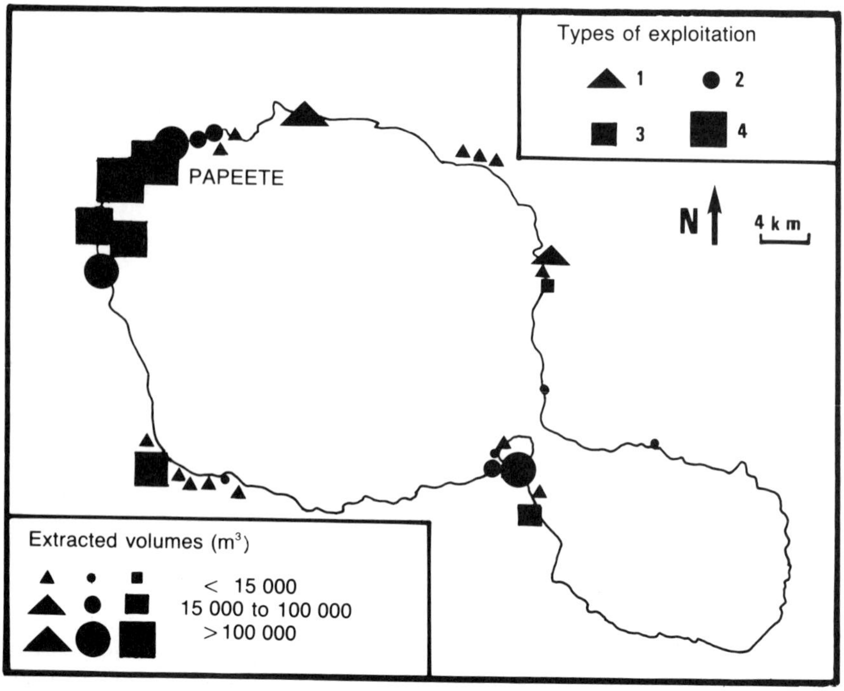

Fig. 149 Reef dredging sites at Tahiti. After Gabrie, Porcher and Masson (1985). 1: channels and small harbours. 2: coral mining. 3: coastal development. 4: extensive coastal development grouping several sites

al., 1985) (Fig. 149), the main source of construction material, mainly for road and urban development, has always been the coral reefs, which have been exploited by dredging. The extraction of these materials has led to a multiplication of exploited sites, mainly in the vicinity of Papeete, Tahiti, but also around Moorea and the Leeward Islands such as Raiatea. In Tahiti alone there were 36 sites of reef mining in 1985. Extraction was generally by hydraulic dredges working along dredging paths. The results are a series of excavations with a chaotic aspect and an irregular geometry, completely disturbing the structure of the reefs. In the lagoon of Grand Cul de Sac Marin on Guadeloupe Island in the French Carribean (Galzin, 1981), dredging of coral sand has led to a considerable proliferation of halophilous bacteria in waters over the dredged areas, related to the suspension of fine particles: an outcome is the disappearance of 20 of the 29 fish species living around the reefs. At Phuket Island, Thailand (Chansang *et al*., 1981), the coastal fringe and nearshore area have been mined for tin for 70 years by dredging, where it has resulted in heavy damage, the lower parts of neighbouring slopes being covered with fine sediment 2.5 centimetres thick. Nickel laterites have been exploited in many coral reef areas (Carey, 1981) in Central America and the Caribbean, Papua New Guinea, Indonesia, Queensland, Philippines. The effects on reefs are generally little known; but investigations in New Caledonia (Bird *et al.*, 1984), where opencast mining for nickel is considerable, have shown the resulting deposition of red muds near the mouths of several rivers and the modification of coastal sedimentation affecting some fringing reefs; other New Caledonian coral reefs lie generally farther from the mainland, and have so far escaped the effect of clays washed down from hilltop mines, but parts of the lagoon are not so secure.

When reefs exist in front of expanding cities, urbanization may encroach upon them. This has recently happened at Jeddah, Saudi Arabia, where the author has observed in 1981 the deterioration of the fringing reef that in the nineteen fifties was living all along that part of the Red Sea coast (Guilcher, 1955). From 1950 to 1980, Jeddah grew from 50,000 to more a million inhabitants. A part of its expansion occurs on the coastal side, to the detriment of the former boat channel, upon which a maritime 'corniche' (sic) of wealthy residences and palaces with swimming pools has been built. The outer edge of the reef has become adjacent to the built-up area, in front of a wall made of huge blocks extracted from the crystalline basement and from the fossil reef existing farther inland: in such a position corals are unlikely to survive. In front of the downtown skyscrapers, the reclamations are partly used for car parking: and there too, the reef has been destroyed.

Another danger derives from soil erosion and use of pesticides for agriculture on lands behind coral reefs. Such effects are common as results of tremendous growth and modernisation of agriculture in developing countries. A typical example has been reported from Mayotte Island, Western Indian Ocean (Guilcher, 1985a), a volcanic island surrounded by a large barrier reef. Investigations made in the lagoon in 1959 had shown that the calcium carbonate content of sediments was higher than 95 per cent in the outer and middle parts; but more recent sampling in 1983 by Thomassin's team pointed, however (Fig. 150), to a decrease in calcium carbonate in the eastern area from 67–99 per cent to 50–71 per cent. At the same time, a whitening and a massive death of corals, and a burying of corals under mud, were observed in the same area (Faure *et al.*, 1984). This increase in terrigenous supply to the lagoon is undoubtedly related to the rapid increase of Mayotte population, which grew from 23,364 inhabitants in 1958 to 52,035 in 1980 (62 to 139 per square kilometre), following a moderate increase in previous years, and led to an extension of the track network, cultivation on steep slopes, and consequent very heavy soil erosion (Figs. 151 and 152).

In the Ryu Kyus, Japan (Muzik, 1985), intensive development for industry and tourism has had damaging effects on the marine environment. 'Natural vegetation has been intensively removed for dams and pineapple fields. River banks and shorelines have been paved with cement, leading to extreme loss of topsoil with every rainfall. Terrigenous sediments are measurably accumulating on the beaches, in lagoons, and even on the reef slope. Living coral cover, measured in % in 1972 and in 1984 in 67 stations around Okinawa Island, has passed in places from 50–90 to 2, from 20–70 to 1–5, from 50–100 to 2, from

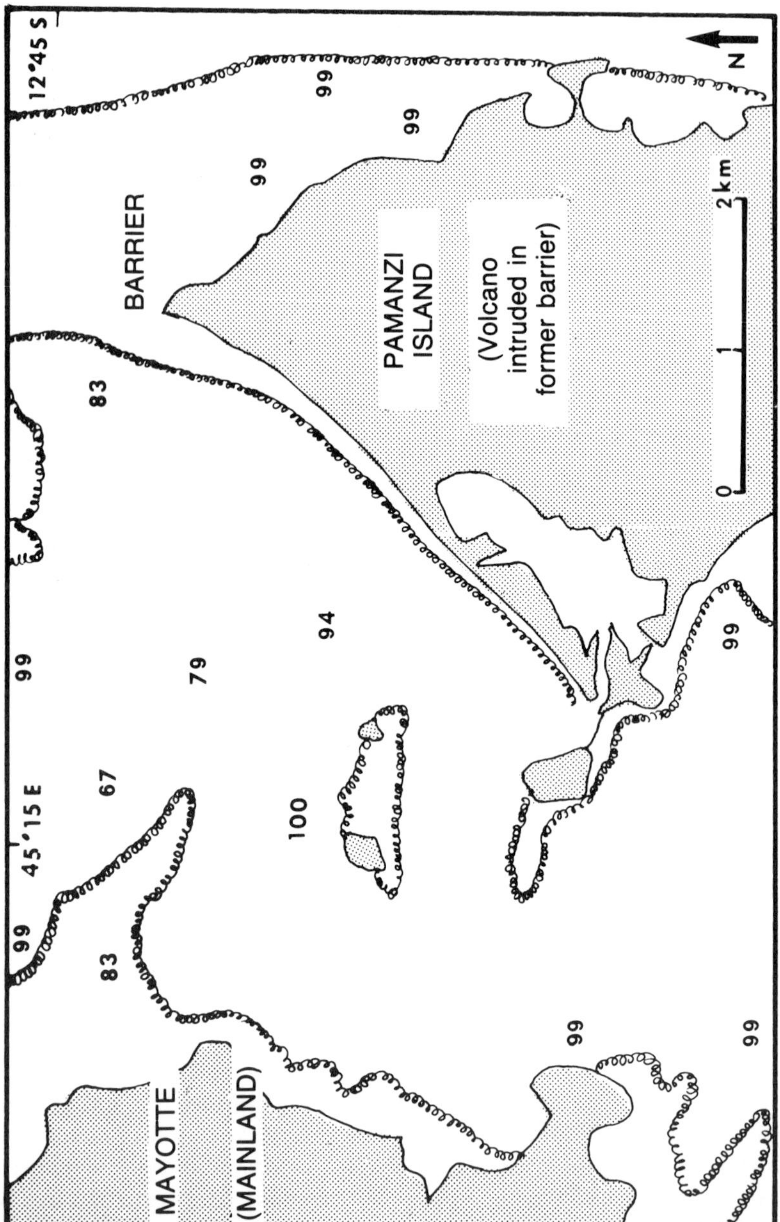
12°45 S
45°15 E
N
BARRIER
PAMANZI
ISLAND
(Volcano
intruded in
former barrier)
MAYOTTE
(MAINLAND)
0
1
2 km
99
99
99
99
99
99
99
99
83
83
67
79
94
100

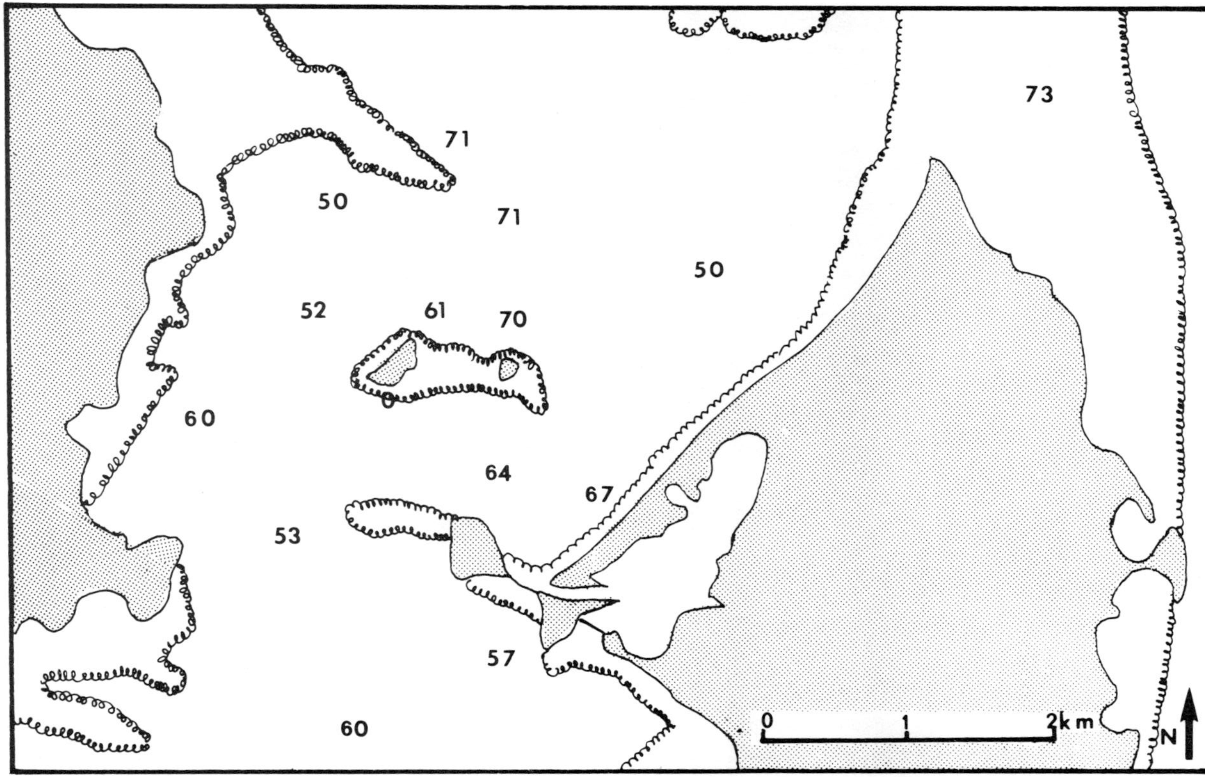

Fig. 150 Eastern part of Mayotte lagoon, Indian Ocean. Decrease from 1959 (upper figure) to 1983 (lower figure) of $CaCO_3$ percentages in sediments, as a result of terrigenous silting through hillslope erosion. Data from Berthois in Guilcher *et al*. (1965a), Gout (unpublished), Arnoux *et al*. (1983; see Guilcher, 1985).

Fig. 151 Hillslope erosion on a volcanic mountain in Mayotte is a cause of siltation in the lagoon (Photo A. Guilcher)

Fig. 152 A microatoll in Mayotte lagoon, with muddy sedimentation in the middle of the colony (Photo A. Guilcher)

50–80 to 2, etc. Not only the amount of living coral cover has decreased, but also the number of species (for example, from 78 to 18 in one station)'. Although original devastation of the coral reefs can be attributed to extensive predation by *Acanthaster planci*, there can be no doubt that the problems of coral reef survival have been compounded by man, as a result of agriculture, tourism and industry.

Discharge of sewage is also responsible for large losses on coral reefs. Along the northwestern coast of Tahiti, where 74 per cent of the island population lives, observations have been made by the Laboratory of Environmental Study and Survey in Tahiti (Fraizier *et al*., 1985), which reported that 'various forms of chemical aggression have been identified and quantified from analysis of waters and/or sediments: detergents, hydrocarbons, pesticides, metals'. Three rivers flowing to the lagoon have particularly contributed metals, and faecal germs from urbanization, Papeete harbour, and local industries are responsible for further damage.

Oahu in Hawaii is another island with coral reefs where the impact of tourism by sewage discharge may be very large. It has been studied in Kaneohe Bay for several years by biologists, geographers and engineers (Banner, 1974; Maragos *et al*., 1985). This bay, which is rather confined, with a poor circulatory pattern in its southeastern part, had flourishing coral reefs before 1939. Since the end of the war, the watershed about the bay has been subject to urbanization, the population increasing almost ten-fold. The result was that the waters of the bay were marked by eutrophication by massive sewage discharge, increased run-off from the watershed in storms, and increased siltation from raw lands left in the course of development of new subdivisions: so that, by 1974, no coral was left growing in one third of the bay, in another one third almost all coral had been overgrown by green algae, and only in the last third (northwestern corner), where the circulation was better, the coral reefs were reasonably intact.

For these two places, both very typical, the studies and works undertaken to counteract the pollution will be mentioned in the second section of this chapter.

Collection of specimens by and for visitors is another cause of degradation. Corals and marine shells have attracted men for a very long time. African populations have used cowries as money since remote periods, oyster shells from lagoons were and are still used as collars in Papua New Guinea, and paintings of the sixteenth century show the appreciation of marine shells by Italian ladies at that epoch. But the damage to reefs remained minor until the recent development of tourism allowed crowds of American, European and Japanese people to visit the reefs, walk on them, dive, and collect fine specimens. The habit of selling shell collars for farewell at airports, in Tahiti for example (several dozens of collars for those who have many friends in the island), is now a profitable industry for local people, who collect the material from the reefs of the surrounding islands. And, for amateurs not able to collect their own specimens from reefs, shops around the world are largely supplied by contractors.

The international trade in ornamental corals and shells has thus known a considerable increase in the 1970s (Wells, 1981). Taiwan and Japan have been the main centres of the precious coral industry; Philippines were the main exporter of stony corals; Philippines, Mexico, Haiti have been among the major suppliers of ornamental shells; the main consumers being the US, Japan and several European countries. From the early 1970s, Japanese and Taiwan coral fishing fleets have been organized, operating in waters off Haiti, the Philippines and Australia as their own waters were rapidly depleted. Exports from the Philippines grew from 631 tons in 1976 to 962 tons in 1978; for Taiwan, from 26 to 206 in the same period. The main shells taken are *Strombus gigas, Cyprea tigris, Nautilius pompilius*, and species of Volutidae, Conidae. It has been mentioned in Chapter two that large collections of *Charonia tritonis* may be responsible for the outbreak of *Acanthaster planci* on the Great Barrier Reef and at Mauritius, and hence the destruction of coral colonies. In the Maldives (Munch-Petersen, 1985), small scale collection of precious corals and shells had not damaged the reefs over a long period, but with recent big increases depletion has become disastrous there too.

In many areas, there is a combination of the abovementioned causes of destruction, which makes the situation quite dangerous. Martinique Island in the Caribbean, for example (Laborel and Bouchon, 1986), was an area where 45 coral species had been identified. There has been damage by heavy siltation, especially in Fort de France Bay and on the southwestern coast, where it derives from sewage tied to urbanization, industrial development, deforestation of hillsides, and road works in the vicinity of the coast. Deposition of mud (Fig. 153) has spread from the centre of the colonies, particularly the small foliated ones, while the meandroid forms are not so easily invaded. Several areas, that were still very rich a few decades ago, are now buried under a general muddy mantle, where only the shape of the former colonies still appear as ghosts. Similarly in Thailand (Sudara, 1981), mining activities, dynamite fishing, bottom trawling, coral collecting, industrial and domestic pollution occur at the same time. Pollution and destructive fishing techniques have also combined in Tonga (Chesher, 1985), where corals are deliberately broken by women and children during fishing activities and for tourists. In Fiji, there has been overfishing, owing to population pressure, and impacts from tourism development and pollution in urban areas (Dahl, 1985). In American Samoa, the best documented country in the South Pacific, a considerable decline in reef quality has been observed, and very serious degradations have been reported around Pago Pago harbour (Dahl, 1985). In Ponape in the Carolines (Holthus, 1985), destruction of habitat from reef flat dredging, terrestrial sedimentation from construction activities, degradation from sand extraction, increased fishing pressure, decreased water quality have all occurred. As a whole, 'most of the accessible reefs in Polynesia are in various stages of decline' (Dahl, 1985). Also Caribbean reefs (Rogers, 1985) have significantly deteriorated since 1975, damaged by a combination of large

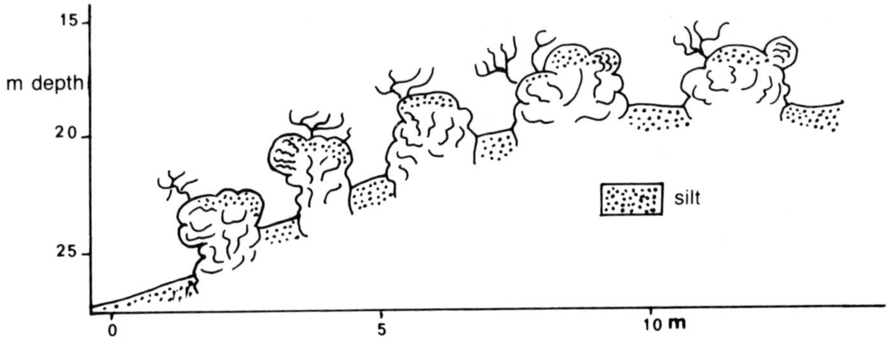

Fig. 153 Siltation in progress on reefs in Fort de France Bay, Martinique, Caribbean. After Laborel and Bouchon (1986)

hurricanes and human activities. Runoff following deforestation of hillsides is a major problem in Guadeloupe, Costa Rica and the Virgin Islands. Careless anchoring and boat grounding are other causes of destruction, as are dredging, use of explosives in fishing, coral collecting, agricultural chemicals. Fig. 154 shows the distribution of these deteriorations in the Caribbean.

2. EFFORTS AGAINST DEGRADATION

The main efforts in this way have been on the Great Barrier Reef of Queensland, which include the widest area of coral reefs in the world. They

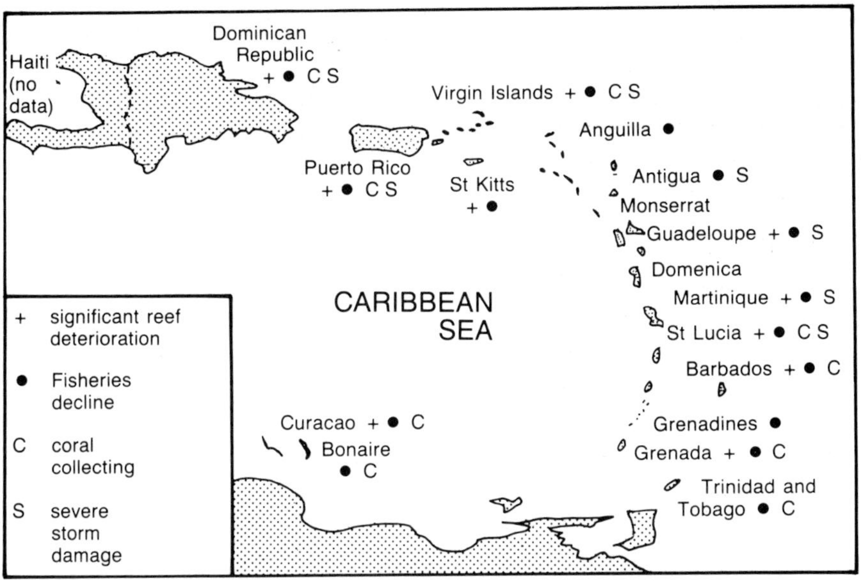

Fig. 154 Reef degradation in Eastern Caribbean. Data from Rogers (1985)

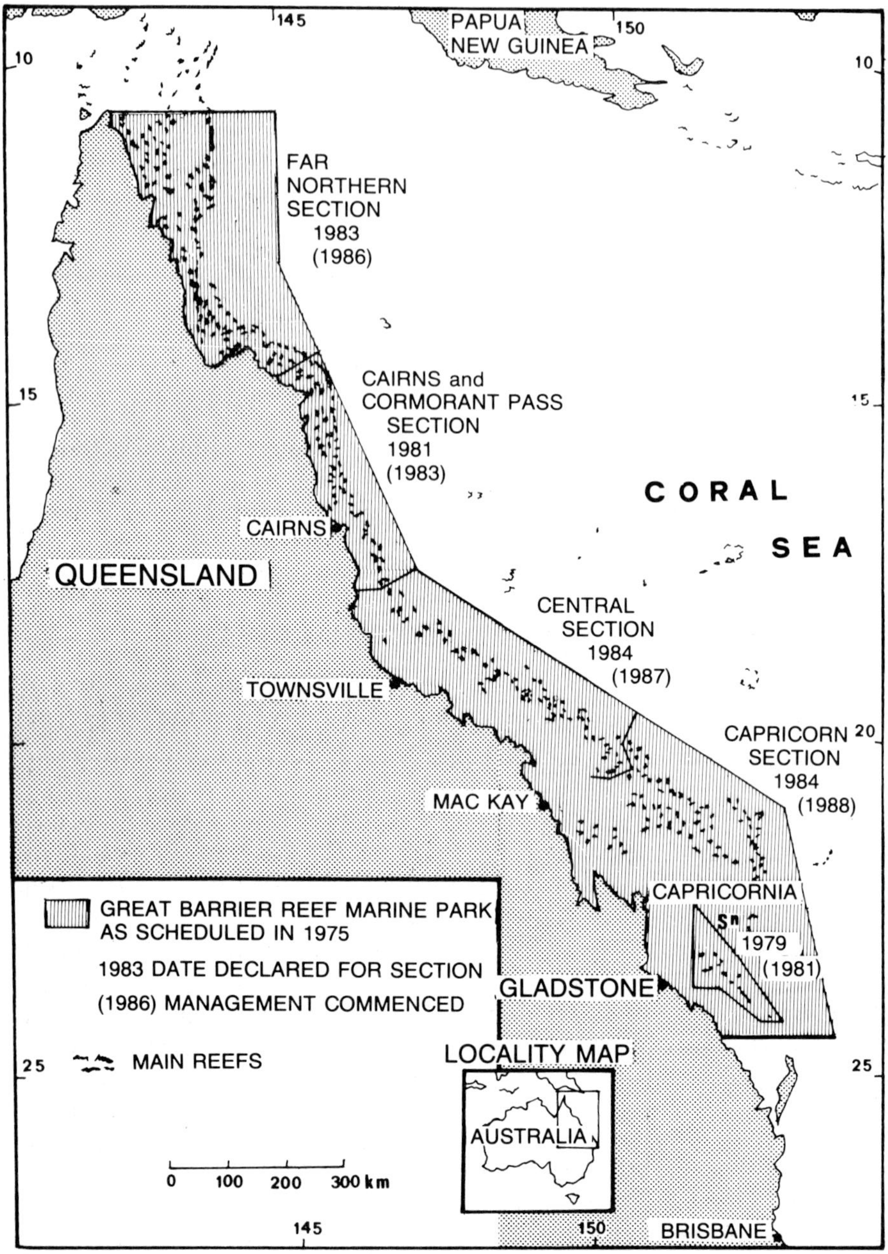

Fig. 155 The Great Barrier Reef Marine Park. After Woodley (1985)

Table 1 Great Barrier Reef Marine Park Status (after Woodley 1985)

Section	Approximate area (km^2)	Date declared	Management commenced
Capricornia	11,800	17.10.79	1.7.81
Cairns and Cormorant Pass	35,000	19.11.81 and 21.10.81	11.11.83
Far Northern	83,000	31.8.83	proposed 1.1.86
Central	77,000	15.10.84	proposed 1.1.87
Capricorn	137,000	15.10.84	proposed 1.1.88
TOTAL	343,800 (2476 reefs)		

led to the creation of the Great Barrier Reef Marine Park (GBRMP), which is expected to be completed and under management in 1988 over a toal area of 343,800 km^2 (Woodley, 1985; Dinesen, 1985; Gilmour and Craik, 1985).

The creation of this marine park was decided by the passing of the GBRMP Act by the Australian Government in 1975. It was preceded by many small national parks of Queensland, such as those listed for the northern sector by Saunders (in GBRMP Authority, 1978, p. 90–96). This huge area has been divided into five sectors (Fig. 155), the areas covered and dates of declaration and commencement of managing being shown in Table 1.

'Firstly, a framework for management is constructed through a zoning process which partitions the uses of the Marine Park. This process is analogous to methods developed for town and country planning and similar in some respects to techniques used in the British National Park system. Zoning is complemented by site-specific management plans. Further controls are implemented through the use of permits on particular uses and by periodic closures of particular reefs and surrounding waters. Secondly, the management of reef users within the zoning plan is implemented on a day-to-day basis through a combination of education, surveillance and enforcement . . . Public participation is undertaken not only because it is a legal requirement but also because it has been found to be essential to the successful preparation of zoning plans'. This participation is expected both prior and after the preparation of the draft zoning plan.

As a whole, four specific objectives have been defined related to the effectiveness of zoning plans, health of the reef, impact of the zoning plan on the reef, on users and on the local economy.

In Capricornia, the first and smallest section to be zoned for management, 'some 30 reefs and shoals are included, plus a number of vegetated coral cays associated with 13 of the reefs, the majority of these islands being already Queensland National Parks. The day-to-day management is carried out by the Queensland National Parks and Wildlife Service, under a special arrangement whereby the Commonwealth Government provides initial capital funding, and 50 per cent of operating costs'. The management has been in operation since 1982. Education/information activities are widely included to increase public awareness of the marine park concept. Surveillance uses aerial and surface patrols. Management includes deterrence and detection of infringements, monitoring of natural resources and control of allowed activities.

In the Cairns and Cormorant Pass sections, the second ones to be organized, the allowed uses are different according to areas: in some places, most activities are accepted including fishing, shipping, defence operations, collecting and research (both requiring permits), while in others only scientific research and traditional hunting and fishing are admitted with permits; recovery of minerals (except for research) and spearfishing commercially or with SCUBA are banned in all zones.

As to the commercial coral collecting on the Great Barrier Reef (Oliver and McGinnity, 1985), approximately 45 tonnes of scleractinian coral are harvested annually, and almost exclusively sold domestically as curios or specimens for marine aquaria. Collecting is restricted to licensed areas of reef fronts, each approximately 400 metres long. In 1983 there were only 12 active collectors, two of them accounting for over 60 per cent of the annual harvest. *Pocillopora damicornis* makes up over 70 per cent of the total amount collected. So far, exports amount to less than 150 kilograms per year (but had been larger in the 1970s as said before in this chapter). It might be, however, that the ban of coral exports decided by the Philippines (see later in this chapter) would result in enquiries in Australia by large scale American and European importers: in which case the Marine Park Authorities will have to monitor even more closely the coral harvest. In any case, the absence of a strong social or economic pressure in Australia for an expansion of the coral trade is certainly a very favourable factor for the wealth of the Great Barrier Reef.

Beside these Australian achievements, Salvat (1981) listed 16 other countries who had at that time 55 coral parks and reserves: Israel (1), Djibouti (2), Kenya (3), Malaysia (3), Indonesia (1), Philippines (14), Japan (11), Palau (1), American Samoa (1), Tonga (3), French Polynesia (2), USA (8), Mexico (2), Colombia (1), Costa Rica (1), Brazil (1).

These parks were generally small, but a special mention must be made of those created in the Philippines, the host country of the 4th International Reef Symposium, which has paid a special attention to his reefs, subject to various stresses, perturbation and overexploitation. A protection of sites has been made by creation of parks (Castaneda and Midat, 1981). But management on conventional modes appeared difficult if not impossible: so that a formula of municipal coral reef park was conceptualized. This system consists of the creation of small protected areas by local governments, which are included in the national scheme for marine parks envisaged for the whole country. The brief description of two such parks by Castaneda is interesting. On the other hand, the protective management of coral reef fishes is also under study in the Philippines at a station on steep outer slope of Sulimon Island, in connection with the very high fishing pressure on most reefs of that country (Russ, 1985).

Another point concerning the Philippines is that, after the very large exports of corals in the nineteen seventies (see above), a total ban on coral collection was considered necessary until more investigations of population biology. But, since the collecting industry employed over 1000 individuals, and brought in millions of dollars in export earnings, the ban has been difficult to enforce (Oliver and McGinnity, 1985).

Among other countries, the efforts made in the United States are noteworthy, the comparatively small area covered by coral reefs being very carefully managed. In Florida, where total prohibition of commercial coral collecting has been in force in state waters since 1974 (Oliver *et al*., 1985), two marine sanctuaries exist on reefs at Key Largo and Looe Key (Curtis, 1985; Halas, 1985). Since the anchors of boats greatly damage fragile corals, a system of mooring buoys has been introduced at Key Largo, where the number of visitors is large, and has proved effective, other Caribbean and Pacific countries have sought advice on setting up such buoys to conserve their reefs. Also at Key Largo, the grounding of a small freighter in 1984 caused substantial damage: coral colonies abraded or overturned, reef crushed and covered with coralline debris. A multifaced program for reef recovery was developed and the results will be synthesized into a management document that discusses the extent of damages associated with ship grounding and identifies strategies for reef recovery following such a disturbance. The wreck may thus be used as an experiment useful for other areas.

In Kaneohe Bay on Oahu, Hawaii (Maragos *et al*., 1985), where damage through sewage discharges has been previously described here, pressure from the public and scientific community compelled the local government and military to terminate large sewage discharges by 1978 in the southeast lagoon, which was most threatened. New surveys in 1983 revealed a remarkable recovery of corals in the southern and middle lagoon, and continued coral abundance in the northern lagoon where no decline had been previously observed. Green algae, on the contrary, which had proliferated through eutrophication, declined greatly. Here too, the results of studies may be applicable elsewhere in management.

In Tahiti the abovementioned observations on lagoon pollution (Fraizier *et al*., 1985) have led to recommendations for improvement. Eighteen investigated stations have been classified according to the quality of the environment, in order to be used as guidelines for further work. The process is, however, ten years behind relative to Oahu. Concerning the dredging of coral material around Tahiti and Moorea (Gabrie *et al*., 1985), it is planned to limit the number of coral reef areas to be dredged, to restore damaged sites (restoration already being undertaken at Moorea), to use improved techniques in places where dredging continues (techniques already tested, that limit the flow of fine sediment), and develop the use of substitute material, such as basaltic rocks and alluvial deposits exploited in closed excavations on Tahiti Island.

Achievements for reef protection have been recorded in the Caribbean, especially at Barbados (Cotter, 1981), where works were undertaken in 1979, and a reserve 2.5 × 1 kilometres opened in 1980 for environmental monitoring, local people and tourist education concerning conservation of coral reef ecosystems. In the Virgin Islands (Rogers, 1985), the United States manage a national park where the role of biosphere reserves in conserving the marine resources is explored. There too an emphasis is made on local people and their involvement in conservation and management.

In New Caledonia (Bird *et al.*, 1984), in spite of the fact that the reefs themselves have not yet been much affected by opencast hilltop mining of nickel ore, the damage to vegetation, slope and valley landscape has led the Association Néo-Calédonienne pour la Sauvegarde de la Nature to promote an interest in nature conservation and landscape protection, especially to preserve the remaining *Araucaria* forests. It is not too late to start restoration, since in areas where mining ceased three or four decades ago, vegetation has revived on the river banks and the red clay has been washed away downstream; it continues, however, to flow towards the river mouths and out into the neighbouring sea, where it increasingly threatens the coral reefs. But it may be hoped that efforts in managing the inland landscapes will help to prevent its deleterious impacts.

At Aldabra, plans for a military base were finally abandoned during the economic crisis of 1967, for financial rather than scientific reasons; however, a large campaign in England, supported by the Royal Society, added its weight for the preservation of the area (Stoddart, 1968). A big naval American base has since been created in the same ocean at Diego Garcia Atoll. It is probably impossible to reconcile all imperatives in coral areas; it was probably more important to preserve nature at Aldabra than at Diego Garcia since its fauna showed unique features.

3. PERSPECTIVES, HINDRANCES AND PHILOSOPHY

A crucial point in the improvement of reef management is the awareness of officials and the general population of the importance of coral reefs and their conservation. Such an awareness is comparatively easy to obtain in developed countries, but it is necessary that it grows in developing countries, where a very large number of coral reefs are found. It has already been observed in states such as the Sudan (Schroeder, 1981; Nasr, 1985), which has a very fine collection of reefs on its coast. The Sudan Marine Conservation Committee, comprising all departments which deal with the sea, has since 1980 been the driving force in marine conservational activities in that country. It has proposed projects such as beaches for recreation, protected diving sites, and the Port Sudan Marine National Park. A series of talks, accompanied by slides and movies about the conservation of marine environment, were given to a wide spectrum of Sudanese communities in Arabic and English languages. The reaction of the audience showed that the message was well received. Public education was also developed by the direct observation of suitable sites, with the help of invited foreign marine scientists. Support from ICUN and the World Wildlife Fund was obtained. A visit of the Duke of Edinburgh to Sanganeb Atoll and reefs contributed to the spread of the idea of establishing a marine national park in that offshore area. Foreign financial support was obtained to strengthen the national marine conservational institutions, and to repair the Sudanese survey vessel for oceanographic investigations. So that within a few years the basic ideas were promoted, and public understanding has been growing.

Various perspectives have appeared in other developing countries, such as Indonesia (Soegiarto, 1981), where more than 40 potential sites have been under study for the establishment of a marine park system, with assistance by the World Wildlife Fund, FAO, UNESCO, and UNDP. Such assistance is quite necessary in the Third World, since the lack of money in these countries is obviously a major problem.

On the other hand, the specificity of developing nations in this respect has been well stressed upon for Ponape Island, in the Carolines, where Holthus (1985) noted that 'the luxury of delaying or removing reef resources from use while studies are conducted or preserves are set aside is unrealistic. Increasing numbers of indigenous people depend upon coral reefs for their livelihood, and resources for short-term economic gains. Reef management must reflect these economic, social and political realities in order to remain viable and achieve its goals'. He thus pointed out the main problem in coral reef management in the Third World, and his remarks help us to understand what has happened in Sri Lanka, for example, where (De Silva, 1985) legislation is adequate to prevent the degradation of coral reefs, under several acts and ordinances, but 'lack of enforcement and political considerations and priorities of a developing nation have rendered this legislation ineffective. A few reefs which have been declared reserves or sanctuaries have not functioned as such due to lack of follow-up regulations and a clear policy on an enforcement authority' (in so far as such a policy would be feasible in this social environment). The case of Sudan has demonstrated that education is an essential tool, but education cannot solve everything. Local scientists and administrators who have such an education in charge must learn the constraints of the society in which they work, and modify accordingly their goals.

Even in developed countries, rational management of coral reefs probably cannot obey exactly the same rules everywhere. Bradbury *et al.* (1985) maintain that good policy in this field 'should be qualitative, local, adaptative, and concerned with the whole ecosystem'. These principles derive from the fact that coral reef ecosystems are characterized by their spatial and temporal heterogeneity. 'Beware the reef manager who claims to have *the* answer'. It is perhaps not surprising that such a statement has been expressed by Australian scientists, who have worked in the huge area of the Great Barrier Reef, where the environments are necessarily heterogeneous over 14 degrees of latitude from Torres Strait to Gladstone. Several years of precise management and planning in such diversified reefs may be considered as an insurance of the value of this opinion.

A seminar held on coral reef fisheries during the Tahiti Congress (Munro *et al*., 1985) has similarly concluded that 'a given management plan will vary from country to country and between areas within a country'; but in the various solutions adopted 'it cannot be overemphasized that conservation of the reef environment is an essential prerequisite to managing the fisheries'. The main administrative management options should be to restrict harvests with size limits, catch quotas and seasonal closures, to create permanent closed reserves, and selective fishing.

Coordination between countries which have reefs is also necessary. Attempts on the international scale have been made, particularly in the South Pacific Regional Environment Programme (SPREP), implemented by the South Pacific Commission in Nouméa, New Caledonia, under the guidance of a coordinating group comprising representatives of several international organizations (Zucker, 1985). Meetings have been held to finalize a convention on environmental protection of coral and lagoon ecosystems.

These considerations will perhaps seem to have extended outside the scope of coral reef geomorphology. But the shape and structure of the coral reefs are in the dependance of all their living environment. Molluscs and fishes need suitable water and sedimentary conditions to thrive, and these are closely associated with the reefs. Coral gardens would be dull without schools of fish swimming to and fro and grazing the colonies. Is there a fatal trend towards the impoverishment of the reefs through an ineluctable increase of human pressure? Or may we still hope, as Dahl and Pichon did when opening the symposium on protection and conservation of the reef environment at Tahiti in 1985, that 'a concerted effort by everyone interested in coral reefs can help to stem the tide of reef destruction'? The future will give the answer.

Bibliography

ABBREVIATIONS

ARB: Atoll Research Bulletin, Washington.
BRGM: Bureau de Recherches Géologiques et Minières, Orléans. Title of vol. 89: Second Symposium International sur les Coraux et les Récifs Coralliens Fossiles.
CNRS: Centre National de la Recherche Scientifique, Paris.
CP: Cahiers du Pacifique, Paris.
CR: Coral Reefs, Springer, Heidelberg.
GSAB: Geological Society of America Bulletin.
ICRS or ICRC: International Coral Reef Symposium, or Congress (First: Mandapam Camp, India; Second: Brisbane; Third: Miami; Fourth: Manila; Fifth: Tahiti).
ORSTOM: Office de la Recherche Scientifique et Technique Outre-Mer, Paris.
PTRS: Philosophical Transactions of the Royal Society, London.
RCNC: Expédition Française sur les Récifs Coralliens de la Nouvelle-Calédonie, 1960–1963, Fondation Singer–Polignac, Paris.

Adey, W. H. (1975) The algal ridges and coral reefs of St Croix: their structure and Holocene development. *ARB*, **187**, 67 pp.

Adey, W. H., Adey, P. J., Burke, R. and Kaufman, L. (1977) The Holocene reef systems of Eastern Martinique, French West Indies. *ARB*, **218**, 40 pp.

Adey, W. H. and Burke, R. (1976) Holocene bioherm (algal ridges and bank-barrier reefs) of the Eastern Caribbean. *GSAB*, **87**, 95–109.

Adey, W. H., MacIntyre, I. G., Stuckenrath, R. and Dill, R. F. (1977) Relict barrier reef system off St Croix: its implication with respect to Late Cenozoic coral reef development in the Western Atlantic. *Proc. 3rd ICRS, Miami*, **2**, 15–21.

Agassiz, A. (1903a) The coral reefs of the tropical Pacific. *Mem. Mus. Compar. Zool.*, **28**, 1–410.

Agassiz, A. (1903b) The coral reefs of the Maldives. *Mem. Mus. Compar. Zool.*, **29**, 1–168.

Aissaoui, D. M. and Purser, B. H. (1985) Reef diagenesis: cementation, dolomitisation and dedolomitisation at Mururoa Atoll, French Polynesia. *Proc. 5th ICRC, Tahiti*, **3**, 257–269.

Arkell, W. J. (1935) On the nature, origin and climatic significance of the coral reefs in the vicinity of Oxford. *Quart. J. Geol. Soc. London*, **91**, 77–110.

Armstrong, H. E. *et al.* (1904) *The Atoll of Funafuti*. Roy. Soc. London, 428 pp.

Asano, D. (1942) Coral reefs of the South Sea Islands. *Tohoku Imp. Univ., Geol. and Paleo. Inst.*, Rept. 39, 7–19 (in Japanese: see Burke, 1951 for English summary).

Atlas Okeanov. Vol. 1: Tikhij Okean (1974); *vol. 2: Atlanticeskij i Indijskij Okeany* (1977). Moscow, Ministry of Defence and of the Navy, in fol., 302 and 306 pp. + indexes (in Russian; also sold by Pergamon Press, London, with English gazeteer).

Baines, G. B. K., Beveridge, P. J., and Maragos, J. E. (1974) Storms and island building at Funafuti Atoll, Ellice Islands. *Proc. 2nd ICRS, Brisbane*, 2, 485–496.

Bakus, G. J. (1975) Marine zonation and ecology of Cocos Island, off Central America. *ARB*, **179**, 110 pp.

Balazs, G. H. (1983) Sea turtles and their traditional usage in Tokelau. *ARB*, **279**, 36 pp.

Banner, A. H. (1974) Kaneohe Bay, Hawaii: urban pollution and a coral reef ecosystem. *Proc. 2nd ICRS, Brisbane*, **2**, 685–702.

Bardintzeff, J. M., Brousse, R. and Gachon, A. (1985) Conditions of building of coral reefs on a volcano: Mururoa in Tuamotu and Rurutu in Australes, French Polynesia. *Proc. 5th ICRC, Tahiti*, **6**, 401–405.

Barlow, N. (ed.) (1933) *Charles Darwin's Diary of the Voyage of HMS Beagle*. Cambridge, 541 pp.

Barlow, N. (ed.) (1958) *The Autobiography of Charles Darwin, 1809–1882*. London, 253 pp.

Battistini, R. (1964) *Etude géomorphologique de l'extrême sud de Madagascar*. Thesis Paris (coral reefs: 448–483).

Battistini, R. (1966) La morphologie de l'île Europa. *Mem. Mus. Hist. Nat. Paris, N.S.*, A, **XLI**, 7–18.

Battistini, R. (1969) Le Quaternaire du littoral kenyan entre Mombaza et Malindi. *Bull. Ass. Fr. Et. Quat.*, **6**, 229–238.

Battistini, R. (1972) L'hypothèse de l'absence de hauts stationnements marins quaternaires. *Bull. Ass. Fr. Et. Quat.*, **9**, 75–81.

Battistini, R. (1977) Un atoll à morphologie énigmatique: Farquhar (Seychelles). *Bull. Ass. Géogr. Fr.*, **444**, 195–204.

Battistini, R. (1978) Les récifs coralliens de la Martinique. *Cah. Off. Rech. Sc. Techn. Outremer, ser. Oceanogr.*, **16** (2), 157–177.

Battistini, R., Bourrouilh, F., Chevalier, J. P., Coudray, J., Denizot, M., Faure, G., Fischer, J. C., Guilcher, A., Harmelin-Vivien, M., Jaubert, J., Laborel, J., Montaggioni, L., Masse, J. P., Mauge, L. A., Peyrot-Clausade, M., Pichon, M., Plante, R., Plaziat, J. C., Plessis, Y. B., Richard, G., Salvat, B., Thomassin, B. A., Vasseur, P. and Weydert, P. (1975) Eléments de terminologie récifale indo-pacifique. *Téthys*, **7**(1), 1–111 (French, English, German).

Battistini, R. and Cremers, G. (1972) Geomorphology and vegetation of Iles Glorieuses. *ARB*, **159**, 25 pp.

Battistini, R. and Hinschberger, F. (1985) La morphologie des côtes au vent de Grande-Terre et de Marie-Galante (Guadeloupe). *Bull. Ass. Géogr. Fr.*, **62**, 86–92.

Battistini, R., Hinschberger, F., Hoang, C. T. and Petit, M. (1986) La basse terrasse corallienne pléistocène (Eemien) de la Guadeloupe: morphologie, datations 230 Th/234 U, néotectonique. *Rev. Géomorph. Dyn.*, **35**, 1–10.

Battistini, R. and Jouannic, C. (1979) Recherches sur la géomorphologie de l'atoll Farquhar (Archipel des Seychelles). *ARB*, **230**, 40 pp.

Battistini, R. and Petit, M. (1979) Récifs coralliens, constructions algaires et arrecifes à la Guadeloupe, Marie-Galante et la Désirade. *ARB*, **234**, 81 pp.

Bendukidze, N. S. (1977) Ecology of Malm Reef formation in the Great Caucasus. *Mem. BRGM*, **89**, 313–321.

Benjamin, G. J. (1970) Diving in the blue holes of the Bahamas. *Nat. Geogr. Mag.*, **138**, 347–363.

Berry, P. and Marsh, L. (1985) Scott Reef and the Rowley Shoals, shelf-edge atolls off Northwestern Australia. *Proc. 5th ICRC, Tahiti*, **6**, 317–322.

Bird, E. C. F. (1971) The fringing reefs near Yule Point, North Queensland, *Australian Geogr. Studies*, **9**, 107–115.

Bird, E. C. F., Dubois, J. P. and Iltis, J. A. (1984) The impacts of opencast mining on the rivers and coasts of New Caledonia. *The United Nations Univ.*, **NRTS–25/UNUP–505**, 52 pp.

Bird, E. C. F. and Guilcher, A. (1982) Observations préliminaires sur les récifs frangeants actuels du Kenya et sur les formes littorales associées. *Rev. Géomorph. Dyn.*, **31**, 113–125.

Blanc, J. J., Pérès, J. M. and Picard, J. (1959) Coraux profonds et thanatocoenoses quaternaires en Méditerranée. *Topographie et Géologie des Profondeurs océaniques*, CNRS, Paris, 185–192.

Bloom, A. L. (1967) Pleistocene shorelines: a new test of isostasy. *GSAB*, **78**, 1477–1493.
Bloom, A. L. (1977, second printing 1979) *IGCP Project 61, Atlas of sea-level curves*. IGCP, UNESCO, IUGS, mimeogr., 123 pp., bibl., index.
Blumenstock, D. I. editor (1961) A report on typhoon effects upon Jaluit Atoll. *ARB*, **75**, 105 pp.
Bonatti, E. (1985) Punctiform initiation of seafloor spreading in the Red Sea during transition from a continental to an oceanic rift. *Nature*, **316**, 6023, 4 July, 33–37.
Bourrouilh, F. (1974) Données géomorphologiques sur la région de Fresh Creek, Ile Andros (Bahama). *Mar. Geol.*, **16**, 213–235.
Bourrouilh, F. (1977a) Géomorphologie de qualques atolls dits 'soulevés' du Pacifique Ouest et Sud-Ouest, origine et évolution des formes récifales actuelles. *Mem. BRGM*, **89**, 419–439.
Bourrouilh, F. (1977b) Les particules du Pléistocène bahamien et de sa croûte: sédimentologie et diagenèse. *Sed. geol.*, **17**, 247–284.
Bourrouilh, F. (1982) Géométrie et minéralogie des corps sédimentaires dans une mangrove carbonatée sous l'influence des ouragans, Ile Andros, Bahamas. *Mém. Soc. Géol. Fr.*, new Ser., **144**, 77–92.
Bourrouilh, F., Bizon, G. and Baumann, P. (1976) La plate-forme carbonatée de l'île de Rennell (Salomon, Sud-Ouest Pacifique) est au minimum d'âge miocène moyen. *C. R. Ac. Sci. Paris*, **282 D**, 1921–1924.
Bourrouilh-Le Jan, F. G. and Talandier, J. (1985a) Sédimentation et fracturation de haute énergie en milieu récifal: tsunamis, ouragans et cyclones et leurs effets sur la sédimentologie et la géomorphologie de Rangiroa, Tuamotu. *Mar. Geol.*, **67**, 263–333.
Bourrouilh-Le Jan, F. G., Talandier, J. and Salvat, B. (1985b) Early diagenesis from 6,000 years ago and the geomorphology of atoll rims in the Tuamotu. *Proc. 5th ICRC, Tahiti*, **3**, 235–240.
Bradbury, R. H., Reichelt, R. E. and Green, D. G. (1985a) Policy = F (theory, data, models, tools) rational solutions to coral reef conservation. *Proc. 5th ICRC, Tahiti*, **4**, 247–251.
Bradbury, R. H. *et al.* (10 authors) (1985b). *Acanthaster planci*. *Proc. 5th ICRC, Tahiti*, **5**, 303–332 (symposium: 5 papers).
Braithwaite, C. J. R. (1971) Seychelles reefs: structure and development. *Symp. Zool. Soc. London*, **28**, 39–63.
Braithwaite, C. J. R. (1982) Patterns of accretion of reefs in the Sudanese Red Sea. *Marine Geol.*, **46**, 297–325.
Brousse, R. C. (1985) The age of the islands in the Pacific Ocean: volcanism and coral reef build up. *Proc. 5th ICRC, Tahiti*, **6**, 389–400.
Brousse, R. C., Chevalier, J. P., Denizot, M. and Salvat, B. (1974) Etude géomorphologique des Iles Gambier. *CP*, **18**, 9–119.
Brousse, R. C., Chevalier, J. P., Denizot, M. and Salvat, B. (1978) Etude géomorphologique des Iles Marquises. *CP*, **21**, 9–74.
Buddemeier, R. W. and Holladay, G. (1977) Atoll hydrology: island groundwater characteristics and their relationship to diagenesis. *Proc. 3rd ICRS, Miami*, **2**, 167–173.
Buigues, D. (1985) Principal facies and their distribution at Mururoa Atoll, French Polynesia. *Proc. 5th ICRC, Tahiti*, **3**, 249–255.
Burke, H. W. (1951) Contributions by the Japanese to the study of coral reefs. *U.S. Geol. Survey, Military Geol. Branch*, Memo. for record, 43 pp.
Cairns, S. D. and Stanley, G. D. Jr. (1981) Ahermatypic coral banks: living and fossil counterparts. *Proc. 4th ICRS*, Manila, **1**, 611–618.
Cane, M. A. (1986) El Niño. *Ann. Rev. Earth Planet. Sci.*, **14**, 43–70.
Cane, M. A. and Zebiak, S. E. (1985). A theory for El Niño and the Southern Oscillation. *Science*, **228**, 31 May, 1085–1087.

Carey, J. (1981) Nickel mining and refinery wastes in coral reef environs. *Proc. 4th ICRS, Manila*, **1**, 137–146.

Cary, L. R. (1931) Studies on the coral reefs of Tutuila, American Samoa, with special reference to the Alcyonaria. *Papers Tortugas Lab.*, **27**, 53–98.

Castaneda, P. G. and Midat, R. I. (1981) The municipal coral reef park in the Philippines. *Proc. 4th ICRS, Manila*, **1**, 283–285.

Caye, G. and Thomassin, B. (1967) Note préliminaire à une étude écologique de la levée détritique et du platier friable du Grand Récif de Tuléar. Morphologie et hydrodynamisme. *Rec. Trav. St. Mar. Endoume*, Fasc. 4, sér. suppl., **6**, 25–35.

Chalker, B. E., Barnes, D. J. and Dunlop, W. C. (1987). Light and reef-building corals. *Interdisc. Sci. Rev.*, in press.

Chansang, H., Boonyanate, P. and Charuehinda, M. (1981) Effect of sedimentation from coastal mining on coral reefs on the northwestern coast of Phuket Island, Thailand. *Proc. 4th ICRS, Manila*, **1**, 129–136.

Chappell, J. (1974a) Geology of coral terraces, Huon Peninsula, New Guinea: a study of Quaternary tectonic movements and sea-level changes. *GSAB*, **85**, 553–570.

Chappell, J. (1974b) Late Quaternary glacio- and hydro-isostasy, on a layered earth. *Quat. Res.*, **4**, 429–440.

Chappell, J. (1983) Evidence for smoothly falling sea-level relative to North Queensland, Australia, during the past 6,000 years. *Nature*, **302**, 406–408.

Chappell, J. and Veeh, H. H. (1978) Late Quaternary tectonic movements and sea-level changes at Timor and Atauro Island. *GSAB*, **89**, 356–368.

Chauveau, J. C., Deneufbourg, G. and Sarcia, J. A. (1967) Observations sur l'infrastructure de l'atoll de Mururoa (Archipel des Tuamotu, Pacifique Sud). *C.R. Ac. Sci. Paris*, **D 265**, 1113–1116.

Chesher, R. (1985) Practical problems in coral reef utilization and management: a Tongan case study. *Proc. 5th ICRC, Tahiti*, **4**, 213–217.

Chevalier, J. P. (1968) Géomorphologie de l'île Maré. *RCNC*, **3**, 1–56.

Chevalier, J. P. (1972) Observations sur les chenaux incomplets appelés hoa dans les atolls des Tuamotu. *Proc. 1st ICRS, Mandapam*, 477–488.

Chevalier, J. P. (1977) Aperçu sur la faune corallienne récifale du Néogène. *Mém. BRGM*, **89**, 42–48.

Chevalier, J. P. (1979) La faune corallienne (Scléractiniaires et Hydrocoralliaires) de la Polynésie Française. *CP*, **2**, 129–151.

Chevalier, J. P., Denizot, M., Mougin, J. C., Plessis, Y., Salvat, B. (1968) Etude géomorphologique et bionomique de l'atoll de Mururoa (Tuamotu). *CP*, **12**, 1–144.

Chevalier, J. P., Denizot, M., Ricard, M., Salvat, B., Sournia, A. and Vasseur, P. (1979) Géomorphologie de l'atoll de Takapoto. *J. Soc. Océanistes*, **62**, 9–18.

Chevalier, J. P. and Salvat, B. (1976) Etude géomorphologique de l'atoll fermé de Taiaro (Tuamotu, Polynésie Française). *CP*, **19**, 169–201.

Chivas, A., Chappell, J., Polach, H., Pillans, B. and Flood, P. (1986) Radiocarbon evidence for the timing and rate of island development, beach rock formation and phosphatization at Lady Elliot Island, Queensland, Australia. *Mar. Geol.*, **69**, 273–287.

Chubb, J. (1957) The pattern of some Pacific islands chains. *Geol. Mag.*, **94**, 221–228.

Clark, J. A., Farrell, W. E. and Peltier, W. R. (1978). Global changes in post-glacial sea level: a numerical calculation. *Quat. Res.*, **9**, 265–287.

Cloud, P. E. Jr. (1959) Geology of Saipan, Mariana Islands. Part 4: Submarine topography and shoal-water ecology. *U.S. Geol. Surv., Prof. Paper*, **280 K**, 361–445.

Connor, J. L. and Adey, W. H. (1977) The benthic algal composition, standing crop, and productivity, of a Caribbean algal ridge. *ARB*, **211**, 15 pp.

Copper, P. (1974) Structure and development of Early Paleozoic reefs. *Proc. 2nd ICRS, Brisbane*, **1**, 365–386.

Coral Reef Newsletter, Sci. Com. Coral Reefs, Pacific Sci. Ass., Univ. of Guam (Bibliographies and informations).
Cotter, P. J. (1981) The establishment of the Barbados marine reserve. *Proc. 4th ICRS, Manila*, **1**, 332.
Coudray, J. (1976) Recherches sur le Néogène et le Quaternaire marins de la Nouvelle-Calédonie. *RCNC*, **8**, 300 pp.
Coudray, J. (1977) Sédimentation et diagenèse des formations récifales périphériques à la Nouvelle-Calédonie durant le Quaternaire. *Mém. BRGM*, **89**, 407–418.
Coudray, J., Thomassin, B. A. and Vasseur, P. (1985) Comparative geomorphology of New Caledonia and Mayotte barrier reefs. *Proc. 5th ICRC, Tahiti*, **6**, 427–432.
Cousteau, J. Y. and Diole, P. (1973). *Galapagos, Titicaca, the Blue Holes*. Cassell, London, 304 pp.
Crossland, C. (1907–1915) Reports on the marine biology of the Sudanese Red Sea. XVIII. A physical description of Khor Dongonab, Red Sea. *J. Linnean Soc. Zoology*, **31**, 265–286.
Curtis, C. (1985) Investigating reef recovery following a freighter grounding in the Key Largo National Marine Sanctuary (Florida Keys, USA). *Proc. 5th ICRC, Tahiti*, **6**, 471–476.
Daetwyler, C. C. and Kidwell, A. L. (1959) The Gulf of Batabano, a modern carbonate basin. *Proc. 5th World Petrol. Congr., N.Y.*, **1**, 1–21.
Dahl, A. L. (1985) Status and conservation of South Pacific coral reefs. *Proc. 5th ICRC, Tahiti*, **6**, 509–513.
Dalongeville, R., editor (1984) Le beach rock. Colloque tenu à Lyon en novembre 1983. *Trav. Maison de l'Orient*, **8**, Lyons, 199 pp (17 papers with English abstracts, and general bibliography).
Dalongeville, R. (1986) *Formes de corrrosion et de construction organogène des littoraux actuels calcaires de Méditerranée. Termes de comparaison pris en Mer Rouge*. Thesis Brest, 374 + 156 pp. (extensive bibliography).
Dalongeville, R. and Sanlaville, P. (1981) Les marsas du littoral soudanais de la Mer Rouge. *Bull. Soc. Languedocienne de Géogr.*, **15**, 39–48.
Daly, R. A. (1910) Pleistocene glaciation and the coral reef problem. *Am. J. Sci., ser. 4*, **30**, 297–308.
Daly, R. A. (1934). *The changing world of the Ice Age*. Yale Univ. Press, New Haven, 271 pp.
Daly, R. A. (1948) Coral reefs: a review. *Am. J. Sci.*, **246**, 193–207.
Dana, J. D. (1872) *Corals and coral islands*. New York (2nd ed. 1874; 3rd ed. 1890).
Darwin, C. (1842) *The structure and distribution of coral reefs. Being the first part of the Geology of the voyage of the Beagle under the command of Capt. Fitzroy, RN, during the years 1832 to 1836*. London, Smith Elder & Co, 214 pp (2nd ed. 1874; 3rd ed. New York 1889; French transl. of 2nd ed. by Cosserat, Paris, Baillière, 1878).
Davies, P. J. and Montaggioni, L. (seminar convenors) (1985) Reef growth and sea-level change: the environmental signature. *Proc. 5th ICRC, Tahiti*, **3**, 477–515 (extensive bibliography).
Davies, P. S., Stoddart, D. R. and Sigeo, D. C. (1971) Reef forms of Addu Atoll, Maldive Islands. *Symp. Zool. Soc. London*, **28**, 217–259.
Davis, W. M. (1928) The coral reef problem. *Am. Geogr. Soc., Spec. Publ.*, **9**, 596 pp. (extensive bibliography to that date).
Debenay, J. P. (1987) Sedimentology in the Southwestern lagoon of New Caledonia, SW Pacific. *J. Coastal Res.*, **3**, 77–91.
Delesalle, B. (1985). Mataiva Atoll, Tuamotu Archipelago. *5th ICRC, Tahiti*, **1**, 269–322.
Delesalle, B. *et al*. (1985) Environmental survey of Mataiva Atoll. *ARB*, **286**, 40 pp.
Deneufbourg, G. (1969) Les forages de Mururoa. *CP*, **13**, 47–58.

De Silva, M. W. R. N. (1985) Status of the coral reefs of Sri Lanka. *Proc. 5th ICRC*, Tahiti, **6**, 515–518.
Dietrich, G. (1950) Kontinentale Einflüsse auf Temperatur und Salzgehalt des Ozeanswassers. *Deutsches Hydr. Zeit.*, **3**, 33–39.
Dietz, R. S. (1961) Continent and ocean basin evolution by spreading of the sea floor. *Nature*, **190**, 4779, 854–857.
Dill, R. F. (1977) The blue holes. Geologically significant submerged sink holes and caves off British Honduras and Andros, Bahama Islands. *Proc. 3rd ICRS, Miami*, **2**, 237–243.
Dinesen, Z. D. (1985) Day-to-day management of Capricornia, the first section of the Great Barrier Reef Marine Park. *Proc. 5th ICRC, Tahiti*, **4**, 253–258.
Donaldson, J. A. and Ricketts, B. D. (1979) Beachrock in Proterozoic dolostone of the Belcher Islands, Northwest Territories, Canada. *J. Sedim. Petrol.*, **49**, 287–294.
Dons, C. (1944) Norges korallrev. *Kong. Norske Videnskab. Selsk., Forhandl.*, **17** (1943), 37–82.
Doumenge, F. (1963) L'île de Makatea et ses problèmes. *CP*, **5**, 41–67.
Doumenge, F. (1966) L'Homme dans le Pacifique Sud. *Publ. Soc. Océanistes*, **9**, Musée de l'Homme, Paris, 635 pp.
Doumenge, F. (1983) Déséquilibres hydroclimatiques et catastrophes dans le Pacifique intertropical, Juillet 1982–avril 1983. *Ann. de Géogr.*, **92**, 403–413.
Downing, M. (1985) Coral reef communities in an extreme environment: the Northwestern Arabian Gulf. *Proc. 5th ICRC, Tahiti*, **6**, 343–348.
DSDP (1973) *Initial Reports of the Deep Sea Drilling Project*, Leg 15, San Juan, Puerto Rico, to Cristobal, Panama, Dec. 1970–Feb. 1971. Nat. Sci. Foundat. **XV**, 1137 pp.
Drew, E. A. (1977) A photographic survey down the seaward reef-front of Aldabra Atoll. *ARB*, **193**, 17 pp.
Drew, E. A. and Abel, K. M. (1985) Biology, sedimentology and geography of the vast inter-reefal Halimeda meadows within the Great Barrier Reef Province. *Proc. 5th ICRC, Tahiti*, **5**, 15–20.
Du Dresnay, R. (1977) Le milieu récifal fossile du Jurassique inférieur (Lias) dans le domaine des chaînes atlasiques du Maroc. *Mém. BRGM*, **89**, 296–312.
Dugas, F. and Debenay, J. P. (1980, 1981, 1982). Carte sédimentologique et carte annexe du lagon de Nouvelle-Calédonie à 1:50 000, feuilles La Tontouta, Prony and Nouméa, with notices explicatives, Paris, ORSTOM, 44, 35 and 45 pp.
Durham, J. W. (1947) Corals from the Gulf of California and the North Pacific coast of America. *Geol. Soc. Am., Mem.*, **20**, 68 pp.
Ehrhardt, J. P. (1976) Hydrobiologie du lagon de Clipperton. *CP*, **19**, 89–112.
Emery, K. O. (1956a) Sediments and waters of Persian Gulf. *Am. Ass. Petr. Geol. Bull.*, **40**, 2354–2383.
Emery, K. O. (1956b) Marine geology of Johnston Island and its surrounding shallows, Central Pacific Ocean. *GSAB*, **67**, 1505–1520.
Emery, K. O., Tracey, J. I. Jr. and Ladd, H. S. (1954). Geology of Bikini and nearby atolls. *U.S. Geol. Surv. Prof. Paper*, **260 A**, 265 pp. and charts.
Emiliani, C. (1966) Isotopic paleotemperatures. *Science*, **154**, 851–857.
Emiliani, C. (1970) Pleistocene paleotemperatures. *Science*, **168**, 822–825.
Esson, J., Flower, M. F. J., Strong, D. F., Upton, B. G. J. and Wadsworth, W. I. (1970) Geology of the Comores Archipelago, Western Indian Ocean. *Geol. Mag.*, **107**, 549–557.
Fagonee, I. (1985) The status of the coral reefs of Mauritius: elaboration of a coral reef research and development plan. *Proc. 5th ICRC, Tahiti*, **2**, 127.
Fairbridge, R. W. (1946–1947) Notes on the geomorphology of the Pelsart Group of the Houtman's Abrolhos Islands. *J. Roy. Soc. Western Australia*, **33**, 1–41.
Fairbridge, R. W. (1950) Recent and Pleistocene coral reefs of Australia. *J. of Geol.*, **58**, 330–401.

Fairbridge, R. W. (1961) Eustatic changes in sea level. *Phys. Chem. Earth*, **4**, 99–185.

Faure, G. (1974) Morphology and bionomy of the coral reef discontinuities in Rodriguez Island, Mascarene Archipelago, Indian Ocean. *Proc. 2nd ICRS, Brisbane*, **2**, 161–172.

Faure, G., Guillaume, M., Payri, C., Thomassin, B. A., Van Praet, M. and Vasseur, P. (1984) Sur un phénomène remarquable de blanchiement et de mortalité massive des Madréporaires dans le complexe récifal de l'île Mayotte (SW Océan Indien). *C.R. Ac. Sci. Paris*, **299** (111) 15, 637–642.

Faure, G., Thomassin, B. A. and Vasseur, P. (1981) Reef coral assemblages on the windward slopes in the Noumea lagoon, New Caledonia. *Proc. 4th ICRS, Manila*, **2**, 293–301.

Fedorov, V. V., Rubinshteyn, I. G., Danilov, I. V. and Lanin (1980) Bottom landscapes of Saya de Malha Bank in the Indian Ocean. *Oceanology*, **20** (4), 434–439.

Fisk, D. A. and Done, T. J. (1985) Taxonomic and bathymetric patterns of bleaching in corals, Myrmidon Reef, Queensland. *Proc. 5th ICRC, Tahiti*, **6**, 149–154.

Flexor, J. M. and Martin, L. (1979) Sur l'utilisation des grès coquilliers de la région de Salvador (Brésil) dans la reconstruction des lignes de rivage holocènes. *Proc. 1978 Internat. Symp. Coastal Evol. Quaternary*, Sao Paulo, 343–355.

Flint, D. E., Corwin, G., Dings, M. G., Fuller, W. P., MacNeil, F. S. and Saplis, R. A. (1963) Limestone walls at Okinawa. *GSAB*, **64**, 1247–1260.

Flood, P. G. and Heatwole, H. (1986) Coral cay instability and species-turnover of plants at Swain Reefs, Southern Great Barrier Reef, Australia. *J. Coastal Res.*, **2**, 479–496.

Fosberg, F. R. and Sachet, M. H. (eds) (1953) Handbook of Atoll Research. *ARB*, **17**, 129 pp.

Fosberg, F. R., Sachet, M. H. and Stoddart, D. R. (1983) Henderson Island (Southeastern Polynesia): summary of recent knowledge. *ARB*, **272**, 53 pp.

Fraizier, A., Frank, D., Benente, P., Jaouen, R. and Debiard, J. P. (1985) *Observations on the various forms of pollution of the Tahiti lagoon. Proc. 5th ICRC, Tahiti*, **6**, 445–451.

Frankel, E. (1977) Previous Acanthaster aggregations in the Great Barrier Reef. *Proc. 3rd ICRS, Miami*, **1**, 201–208.

Freeman-Lynde, R. P. and Ryan, W. B. F. (1985) Erosional modification of Bahaman escarpment. *GSAB*, **96**, 481–494.

Fryer, P. and Smoot, N. C. (1985) Processes of seamount subduction in the Mariana and Izu–Bonin trenches. *Mar. Geol.*, **64**, 77–90.

Gabrie, C., Porcher, M. and Masson, M. (1985) Dredging in French Polynesia coral reefs: towards a general policy of resource exploitation and site development. *Proc. 5th ICRC, Tahiti*, **4**, 116–121.

Gachon, A. and Buigues, D. (1985) Volcanic erosion and reef growth phases, atoll of Mururoa, French Polynesia. *Proc. 5th ICRC, Tahiti*, **3**, 185–191.

Galzin, R. (1981) Effects of coral and dredging on fish fauna in the lagoon of Grand Cul-de-Sac Marin, Guadeloupe, French West Indies. *Proc. 4th ICRS, Manila*, **1**, 116–121.

Gardiner, J. S. (1903) *The fauna and geography of the Maldive and Laccadive archipelagoes*. Cambridge Univ. Press, vol. 1, 471 pp.

Gaven, C. M. and Bourrouilh, F. (1981) Géochronologie, sédimentologie et néotectonique des facies récifaux pléistocènes à Maré. *Oceanis*, **7** (4), 347–365.

Garret, P. and Gould, J. (1984) Geology of New Providence Island, Bahamas. *GSAB*, **95**, 209–220.

Gilmour, A. and Craik, W. (1985) A framework for monitoring the Great Barrier Reef Marine Park. *Proc. 5th ICRC, Tahiti*, **4**, 265–270.

Gladfelter, E. H. (1985) Metabolism, calcification and carbon production. 11: Organism level studies. *Proc. 5th ICRC, Tahiti*, **4**, 527–539.

Glynn, P. W. and Stewart, R. H. (1973) Distribution of coral reefs in the Pearl Islands (Gulf of Panama) in relation to thermal conditions. *Limnol. and Oceanogr.*, **18**, 367–379.

Goreau, T. F. (1963) Calcium carbonate deposition by coralline algae and corals in relation to their role as reef builders. *Ann. N.Y. Acad. Sci.*, **109**, 127–167.

Grange, K. R. (1985) Distribution, standing crop, population structure and growth rates of an unexploited resource of black coral in the southern fjords of New Zealand. *Proc. 5th ICRC, Tahiti*, **6**, 445–451.

Gravier, N., Harmelin, J. G., Pichon, M., Thomassin, B., Vasseur, P. and Weydert, P. (1970) Les récifs coralliens de Tuléar (Madagascar): morphologie et bionomie de la pente externe. *C.R. Ac. Sci. Paris*, **270**, 1130–1133.

Great Barrier Reef Marine Park Authority (1978) *Workshop on the Northern sector of the Great Barrier Reef*, Townsville, mimeogr., 462 pp.

Grigg, R. W. (1974) Distribution and abundance of precious corals in Hawaii. *Proc. 2nd ICRS, Brisbane*, **2**, 235–240.

Grigg, R. W. (1982) A threshold for atoll formation. *CR*, **1**, 29–34.

Gross, M. G., Milliman, J. D., Tracey, J. I. Jr. and Ladd, H. S. (1969) Marine geology of Kure and Midway Atolls, Hawaii: a preliminary report. *Pacific Sci.*, **23** (1), 17–25.

Guilcher, A. (1950) Les récifs coralliens: formes et origines. *Information Géogr.*, **14** (5), 183–196 (extensive bibliography).

Guilcher, A. (1954) Caractères du récif-barrière de la côte nord-ouest de Madagascar. *C.R. Somm. Soc. Géol. Fr.*, **(6) 4**, 14, 372–373.

Guilcher, A. (1955) Géomorphologie de l'extrémité septentrionale du Banc Farsan (Mer Rouge). *Ann. Inst. Océanogr.*, **XXX**, 55–100.

Guilcher, A. (1956) Etude géomorphologique des récifs coralliens du Nord-Ouest de Madagascar. *Ann. Inst. Océanogr.*, **XXX111-2**, 65–136.

Guilcher, A. (1958) *Coastal and submarine morphology*. London, Methuen, 274 pp. (French edition 1954).

Guilcher, A. (1959) Les récifs coralliens à petits lagons multiples de la baie Ramanetaka (côte nord-ouest de Madagascar). *Bull. Soc. Géol. Fr.*, **(7) 1**, 337–340.

Guilcher, A. (1961) Le beach rock ou grès de plage. *Ann. de Géogr.*, **70**, 113–125.

Guilcher, A. (1963) Quelques caractères des récifs-barrières et de leurs lagons. *Bull. Ass. Géogr. Fr.*, **314–315**, 2–15.

Guilcher, A. (1965a) Grand Récif Sud, Récifs et Lagon de Tuo, Nouvelle-Calédonie. *RCNC*, **1**, 137–240.

Guilcher, A. (1965b) Coral reefs and lagoons of Mayotte Island, Indian Ocean, and of New Caledonia, Pacific Ocean. *Colston Res. Soc., Proc. 17th Symp. at Bristol*. London, Butterworths, 21–45.

Guilcher, A. (1965c) Questions de morphologie climatique en Mélanésie équatoriale (Nouvelle-Guinée, Nouvelle-Bretagne, Guadalcanal). *Bull. Ass. Géogr. Fr.*, **338–339**, 28–40.

Guilcher, A. (1966) Transport de sédiments par-dessus les bordures d'atolls et les récifs-barrières dans le Pacifique Sud. *2nd Int. Oceanogr. Congr.*, Moscow, Morning Review Lectures (publ. UNESCO Paris), 199–206.

Guilcher, A. (1967a) Sedimentation and sediments in atoll lagoons and behind barrier reefs in coral seas. *Lagunas Costeras, Un Simposio*, Univ. Nat. Autonoma, Mexico, 193–220 (published 1969).

Guilcher, A. (1967b) Les îles Gilbert comparées aux Tuamotus. *J. Soc. Océanistes*, **23**, 101–113.

Guilcher, A. (1968) Les eaux dessalées du Golfe de Guinée. *Norois*, **15**, 133–134.

Guilcher, A. (1969) Pleistocene and Holocene sea-level changes. *Earth Sci. Rev.*, **5**, 69–97.

Guilcher, A. (1971) Mayotte barrier reef and lagoon, Comoro Islands, as compared with other barrier reefs, atolls and lagoons in the World. *Symp. Zool. Soc. London*, **28**, 65–86.

Guilcher, A. (1973) Lord Howe, l'île à récifs coralliens la plus méridionale du monde (Mer de Tasman, 31°30′S, 158°E). *Bull. Ass. Géogr. Fr.*, **405**, 427–437.

Guilcher, A. (1974) Coral reefs of the New Hebrides, Melanesia, with special reference to open-sea, not fringing, reefs. *Proc. 2nd ICRS, Brisbane*, **2**, 523–535.

Guilcher, A. (1976) Double and multiple barrier reefs in the world. *Studia Societatis Scientiarum Torunensis* (Torun, Polonia), **VIII C**, 4–6, 85–99 (Galon Commemorative Volume).

Guilcher, A. (1979a) Les rivages coralliens de l'Est et du Sud de la presqu'île du Sinai. *Ann. de Géogr.*, **88**, 393–418.

Guilcher, A. (1979b) Origin of passes across atoll rims and barrier reefs. *Proc. XIVth Pacific Sci. Congr., Khabarovsk*, Com. CD, 113 (abstract).

Guilcher, A. (1982a) Physiography of the Red Sea coast and coastal waters: a summary of present state of knowledge, with suggestions for further research. *J. Fac. Mar. Sci., Jeddah*, **2**, 27–36.

Guilcher, A. (1982b) Problèmes climatico-océanographiques du désert côtier d'Angola, particulièrement à la Baie des Tigres (16°35′S). *Norois*, **29**, 507–517.

Guilcher, A. (1983) Géomorphologie et utilisation de la côte de la Paraiba (Brésil). *Trav. Doc. Géogr. Tropic.*, CEGET, Bordeaux, **50**, 373–425 (coral reefs: 396–401).

Guilcher, A. (1985a) Nature and human change of sedimentation in lagoons behind barrier reefs in the humid tropics. *Proc. 5th ICRC, Tahiti*, **4**, 207–212.

Guilcher, A. (1985b) Red Sea coasts, in Bird and Schwartz (eds) *The World's Coastline*, New York, 713–717.

Guilcher, A. (1985c) Society Islands, in Bird and Schwartz (eds) *The World's Coastline*, New York, 1023–1026.

Guilcher, A. (1986) Developing carbonate platforms: Southern Gulf of Suez, Northern Red Sea: comment (and reply by Roberts and Murray). *Mar. Geol.*, **69**, 363–366.

Guilcher, A., Berthois, L., Battistini, R. and Fourmanoir, P. (1958) Les récifs coralliens des Iles Radama et de la Baie Ramanetaka (côte nord-ouest de Madagascar). *Mém. Inst. Sci. Madagascar*, F, **11**, 117–199.

Guilcher, A., Berthois, L., Le Calvez, Y., Battistini, R. and Crosnier, A. (1965a) Les Récifs coralliens et le lagon de Mayotte (Archipel des Comores, Océan Indien). *Mém. ORSTOM*, **11**, 210 pp.

Guilcher, A., Berthois, L., Doumenge, F. and Michel, A. (1965b) Les teneurs des sédiments en carbonate de calcium dans les lagons de Maupihaa (Mopelia) et de Bora-Bora (Iles de la Société). *C. R. Ac. Sci. Paris*, **261**, 5177–5179.

Guilcher, A., Berthois, L., Doumenge, F., Michel, A. Saint-Requier (A.) and Arnold (R.) (1969) Les récifs et lagons coralliens de Mopelia et de Bora-Bora (Iles de la Société) et récifs de comparaison: Tahiti, Scilly, Tuamotu Occidentales. *Mém. ORSTOM*, **38**, 105 pp.

Guilcher, A. and Marec, A. (1978) Le récif–barrière et le lagon du Grand Cul-de-Sac Marin (Guadeloupe): géomorphologie et sédiments. *Oceanol. Acta*, **1** (4), 435–444.

Guilcher, A., Medeiros, C. A., De Matos, J. E. and De Oliveira, J. T. (1974) Les restingas (flèches littorales) d'Angola. *Finisterra* (Lisbon), **9**, 171–211.

Guilcher, A., Saint-Requier, A. and Doumenge, F. (1966) Les teneurs des sédiments en carbonate de calcium dans le lagon de Tahiti, et dans les lagons derrière des barrières coralliennes en général. *C. R. Ac. Sci. Paris*, **263 D**, 25–27.

Guppy, H. B. (1888) A criticism of the theory of subsidence as affecting coral reefs. *Scottish Geogr. Mag.*, **4**, 121–137.

Gvirtzman, G., Buchbinder, B., Sneh, A., Nir, Y. and Friedman, G. M. (1977) Morphology of the Red Sea fringing reefs. *Mém. BRGM*, **89**, 480–491.

Gvirtzman, G. and Buchbinder, B. (1978) Recent and Pleistocene coral reefs and coastal sediments of the Gulf of Elat. *10th Int. Congr. Sedimentol., Jerusalem*, Post-Congr. Exc. **4**, 163–191.

Hadding, A. (1941) The Pre-Quaternary sedimentary rocks of Sweden. VI. Reef limestones. *Lunds Univ. Arsskr.*, NF, Avd. 2, Bd. 37, no 10 (*Kungl. Fysiograf. Sällsk. Handl.*, N.F., Bd. 52, no 10), 1–137.

Hajash, A. and Armstrong, R. (1972) Paleomagnetic and radiometric evidence for the age of the Comores Islands, West Central Indian Ocean. *Earth Planet. Sci. Lett.*, **16**, 231–236.

Halas, J. C. (1985) A unique mooring system for reef management in the Key Largo National Marine Sanctuary. *Proc. 5th ICRC, Tahiti*, **4**, 237–242.

Hamada, T. (1977) The Holocene corals of raised reefs of Japan. *Mém. BRGM*, **89**, 389–395.

Hamilton, E. L. (1957) Marine geology of the Southern Hawaiian ridge. *GSAB*, **68**, 1011–1026.

Hanzawa, S. (1940) Micropaleontological studies of drill cores from a deep well in Kita Daito Zima (North Borodino Island). *Jubilee Publ. in Commemor. Prof. H. Yabe's 60th Birthday*, **2**, 755–802.

Harmelin-Vivien, M. (1985) Tikehau Atoll, Tuamotu Archipelago. *Proc. 5th ICRC, Tahiti*, **1**, 211–268.

Harmelin-Vivien, M. and Stoddart, D. R. (convenors) (1985) Symposium on hurricane effects on coral reefs. *Proc. 5th ICRC, Tahiti*, **3**, 315–350 (7 papers).

Hatcher, B. G. (1985) Ecological research at the Houtman's Abrolhos: high latitude reefs of Western Australia. *Proc. 5th ICRC, Tahiti*, **6**, 291–297.

Heidecker, E. (1973) Structural and tectonic features influencing the development of recent coral reefs off Northeastern Queensland. In Jones and Endean (eds) *Biology and Geology of Coral Reefs*, N. York, **1** (1), 273–298.

Hendry, M. D. and Head, S. M. (1985) Late Quaternary sea-level changes and the development of the raised reef/dune sequence at Great Pedro Bluff, SW Jamaica. *Proc. 5th ICRC, Tahiti*, **3**, 119–124.

Henny, R. W., Mercer, J. W. and Zbur, R. T. (1974) Near surface geologic investigations at Eniwetok Atoll. *Proc. 2nd ICRS, Brisbane*, **2**, 615–626.

Hess, H. H. (1962) History of the ocean basins. In *Petrologic Studies, a Vol. to Honour A. F. Buddington, Geol. Soc. Am.*, 599–620.

Hillis-Colinvaux, L. (1974) Productivity of the coral reef alga *Halimeda*. *Proc. 2nd ICRS, Brisbane*, **1**, 595–619.

Hillis-Colinvaux, L. (1985) *Halimeda* and other deep fore-reef algae at Enewetak Atoll. *Proc. 5th ICRC, Tahiti*, **5**, 9–14.

Hine, A. C., Wilber, R. J., Bane, J. M., Neumann, A. C. and Lorensen, K. R. (1981) Offbank transport of carbonate sands along open, leeward bank margins: northern Bahamas. *Mar. Geol.*, **42**, 327–348.

Hirst, A. C. and Hastenrath, S. (1983) Atmosphere–ocean mechanisms of climate anomalies in the Angola–Tropical Atlantic sector. *J. Phys. Oceanogr.*, **13**, 1146–1157.

Hixon, M. A. and Brostoff, W. N. (1981) Fish grazing and community structure of Hawaiian reef algae. *Proc. 4th ICRS, Manila*, **2**, 431–437.

Hodgkin, E. P. (1970) Geomorphology and biological erosion of limestone coasts in Malaysia. *Geol. Soc. Malaysia Bull.*, **3**, 27–51.

Hoffmeister, J. E. (1940) James Dwight Dana's studies of volcanoes and of coral islands. *Proc. Am. Philos. Soc.*, **82**, 5, 721–732.

Hoffmeister, J. E. and Ladd, H. S. (1944) The Antecedent-Platform Theory. *J. of Geol.*, **52**, 388–402 (and Id., *Ibid.*, 1935, **43**, 653–665; 1936, **44**, 74–92).

Hoffmeister, J. E. and Ladd, H. S. (1945) Solution effects on elevated limestone terraces. *GSAB*, **56**, 809–818.

Hoffmeister, J. E., Stockman, K. W. and Multer, H. G. (1967) Miami limestone of Florida and its recent Bahamian counterpart. *GSAB*, **78**, 175–190.

Holthus, P. F. (1985) A reef resource conservation and management plan for Ponape Island, Caroline Archipelago. *Proc. 5th ICRC, Tahiti*, **4**, 231–236.

Hopkins, D. M. (1967) Quaternary marine transgressions in Alaska. In Hopkins (ed) *The Bering Land Bridge*, Stanford Univ. Press, Cal., 47–90.

Hopley, D. (1982) *The Geomorphology of the Great Barrier Reef.* New York, Wiley, 453 pp. (extensive bibliography).

Hopley, D. and Murtha, G. G. (1975) The Quaternary deposits of the Townsville coastal plain. *Dept. Geogr. James Cook Univ. of North Queensland, Monogr. Ser.*, **8**, 30 pp.

Hopley, D. and Pirazzoli, P., chairmen (1985) Late Quaternary and present sea-level changes: magnitude, causes, future applications. *Proc. 5th ICRC, Tahiti*, **3**, 77–78.

Illing, L. V. (1954) Bahaman calcareous sands. *Am. Ass. Petr. Geol. Bull.*, **38**, 1–95.

Jaap, W. (1985) An epidemic Zooxanthellae expulsion during 1983 in the Lower Florida Keys coral reefs. *Proc. 5th ICRC, Tahiti*, **6**, 143–148.

Johnson, D. P., Searle, D. E. and Hopley, D. (1982) Positive relief over buried post-glacial channels, Great Barrier Reef Province, Australia. *Mar. Geol.*, **46**, 149–159.

Kaplin, P. A., Leontyev, O. K. and Medvedev, V. J. *et al.* (1973) *Geography of atolls in the Southwestern Pacific Ocean*. Science Edit., Moscow, 140 pp. (in Russian, no abstract in other language).

Karig, D. E. and Mammerickx, J. (1972) Tectonic framework of the New Hebrides Island Arc. *Mar. Geol.*, **12**, 187–205.

Keating, B. (1985) Submersible observations on the flanks of Johnston Island, Central Pacific Ocean. *Proc. 5th ICRC, Tahiti*, **6**, 413–418.

Kent, D., Opdyke, N. D. and Ewing, M. (1971) Climate change in the North Pacific using ice-rafted detritus as a climatic indicator. *GSAB*, **82**, 2741–2754.

Kidson, C. (1982) Sea level changes in the Holocene. *Quat. Sci. Rev.*, **1** (2), 121–151.

Kiene, W. E. (1985) Biological destruction of experimental coral substrates at Lizard Island, Great Barrier Reef, Australia. *Proc. 5th ICRC, Tahiti*, **5**, 339–344.

Kinsey, D. W. (1985) Metabolism, calcification and carbon production. I. System level studies. *Proc. 5th ICRC, Tahiti*, **4**, 505–526.

Krempf, A. (1927) La forme des récifs coralliens et le régime des vents alternants. *Mém. Trav. Serv. Océanogr. Pêches Indochine*, **2**, 33 pp.

Kudoja, W. M. (1985) The Tanzanian coral reefs at risk. *Proc. 5th ICRC, Tahiti*, **2**, 209.

Kuenen, Ph. H. (1933) Geology of coral reefs. *The Snellius expedition in the eastern part of the Netherlands East Indies, 1929–1930*. vol. V. Geological results, part 2. Utrecht, 126 pp.

Kuenen, Ph. H. (1947) Two problems in marine geology, atolls and canyons. *Verh. Kon. Ned. Akad. Wet., Afd. Natuurk.*, Tweede Sekt., **D. 43**, 3, 69 pp.

Kuenen, Ph. H. (1950) *Marine Geology*. Wiley & Chapman, New York & London (coral reefs: 414–479).

Kühlmann, D. H. H. (1970–1971–1974) Die Korallenriffe Kubas. *Int. Revue Ges. Hydrobiol.*, **55**, 729–756; **56**, 145–199; **59**, 305–325.

Kühlmann, D. H. H. (1974) The coral reefs of Cuba. *Proc. 2nd ICRS, Brisbane*, **2**, 69–83.

Kühlmann, D. H. H. (1975) Charakterisierung der Korallenriffe von Veracruz, Mexiko. *Int. Revue Ges. Hydrobiol.*, **60**, 495–521.

Labeyrie, J., Lalou, C. and Delibrias, G. (1969) Etude des transgressions marines sur l'atoll de Mururoa par la datation des différents niveaux de corail. *CP*, **13**, 59–68.

Laborel, J. (1969) *Les peuplements de Madréporaires des côtes tropicales du Brésil.* Thesis Marseilles, and Ann. Univ. Abidjan, E–11–3, 261 pp.

Laborel, J. (1974) West African reef corals: a hypothesis on their origin. *Proc. 2nd ICRS, Brisbane*, **1**, 425–443.
Laborel, J. and Bouchon, C. (1986) Les peuplements coralliens des côtes de la Martinique. *Ann. Inst. Océanogr.*, **62** (2), 193–237.
Ladd, H. S. and Hoffmeister, J. E. (1945) Geology of Lau, Fiji. *Bernice P. Bishop Mus., Bull.*, **181**, 1–190.
Ladd, H. S., Newman, W. A. and Sohl, N. F. (1974) Darwin Guyot, the Pacific's oldest atoll. *Proc. 2nd ICRS, Brisbane*, **2**, 513–522.
Ladd, H. S., Tracey, J. I. Jr. and Lill, G. G. (1948) Drilling on Bikini Island, Marshall Islands. *Science*, **107**, no 2768, 51–55.
Ladd, H. S., Tracey, J. I. Jr. and Gross, M. G. (1970) Deep drilling on Midway Atoll. *US Geol. Surv. Prof. Paper*, **680 A**, A 1–22.
Ladd, H. S. and Schlanger, S. O. (1960) Drilling operations on Eniwetok Atoll. *US Geol. Surv. Prof. Paper*, **260 Y**, 863–903.
Lalou, C., Labeyrie, J. and Delibrias, G. (1966). Datation des calcaires coralliens de l'atoll de Mururoa (Archipel des Tuamotu) de l'époque actuelle jusqu'à-500 000 ans. *C. R. Ac. Sci. Paris*, **263**, 1946–1949.
Lambeck, K. and Nakada, M. (1985) Holocene fluctuations in sea-level: constraints on mantle viscosity and meltwater sources. *Proc. 5th ICRC, Tahiti*, **3**, 79–84.
Le Campion-Alsumard, T. (1979) Les Cyanophycées endolithes marines. *Oceanol. Acta*, **2**, 143–156.
Lecolle, J. and Bernat, M. (1985) Late Quaternary surrection history of Efate Island, New Hebrides Island Arc (Vanuatu). *Proc. 5th ICRC, Tahiti*, **3**, 179–184.
Lee, M. A. B. (1984) Biogeography and ecology of atoll plants. *Progr. Phys. Geogr.*, **8** (4), 509–522.
Lemoine, P. (1940) Les algues calcaires de la zone néritique. *Soc. Biogéogr. Mém.*, **7**, 75–138.
Leontyev, O. K. and Medvedev, V. S. (1972) Evolution of atolls in the Pacific Ocean. *Priroda*, **9**, 80–87 (in Russian).
Lewis, M. S. (1969) Sedimentary environments and unconsolidated carbonate sediments of the fringing coral reefs of Mahé, Seychelles. *Mar. Geol.*, **7**, 95–127.
Lighty, R. G., MacIntyre, I. G. and Stuckenrath, R. (1982) *Acropora palmata* reef framework: a reliable indicator of sea level in the Western Atlantic for the last 10 000 years. *CR*, **1**, 125–130.
MacNeil, F. S. (1954) The shape of atolls: an inheritance from subaerial erosion forms. *Am. J. Sci.*, **252**, 402–427.
McKee, E. D., Chronic, J. and Leopold, E. B. (1959) Sedimentary belts in lagoon of Kapingamarangi Atoll. *Am. Ass. Petr. Geol. Bull.*, **43**, 501–562.
McNutt, M. and Menard, H. W. (1978) Lithospheric flexure and uplifted atolls. *J. Geophys. Res.*, **83**, 1206–1212.
Mallik, T. K. (1985) Geology of the Kavaratti and Kalpeni Atolls, Lakshadweep, Arabian Sea, Indian Ocean. *Proc. 5th ICRC, Tahiti*, **6**, 419–426.
Maragos, J. E., Evans, C. and Holthus, P. (1985) Reef corals in Kaneohe Bay six years before and after termination of sewage discharges. (Oahu, Hawaiian Archipelago). *Proc. 5th ICRC, Tahiti*, **4**, 189–194.
Marshall, J. F. and Davies, P. J. (1982) Internal structure and Holocene evolution of One Tree Reef, Southern Great Barrier Reef. *CR*, **1**, 21–28.
Marshall, J. F. and Jacobson, G. (1985) Holocene growth of a Mid-Pacific atoll: Tarawa, Kiribati. *CR*, **4**, 11–17.
Martin, L., Flexor, J. M., Blitzkow, D. and Suguio, K. (1985) Geoid change indications along the Brazilian coast during the last 7 000 years. *Proc. 5th ICRC, Tahiti*, **3**, 85–90.
Matsushima, Y. (1979) Littoral molluscan assemblages during the post-glacial Jomon transgression in the Southern Kanto, Japan. *Daiyonki-Kenyu*, **17**, 243–265 (in Japanese with English summary).

Matthews, J. E. and Holcombe, T. L. (1985) Venezuela Basin of the Caribbean Sea. Stratigraphy and sediment distribution. *Mar. Geol.*, **68**, 1–23.

Maugé, L. A., Ségoufin, J., Vernier, E. and Froget, C. (1982) Géomorphologie et origine des bancs du Nord-Est du Canal de Mozambique, Océan Indien Occidental. *Mar. Geol.*, **47**, 37–55.

Menard, H. W. (1964) *Marine Geology of the Pacific*. McGraw Hill, New York, 271 pp.

Meinesz, A., Jaubert, J. and Denizot, M. (1981) Distribution of the algae belonging to the genus *Caulerpa* in French Polynesia (Atoll of Takapoto and Island of Moorea). *Proc. 4th ICRS, Manila*, **2**, 431–437.

Mergner, H. (1967) Ueber den Hydroidenbewuchs einiger Korallenriffe des Roten Meeres. *Zeit. Morph. Oekol. Tiere*, **60**, 35–104.

Mergner, H. (1971) Structure, ecology and zonation of Red Sea reefs. *Symp. Zool. Soc. London*, **28**, 141–161.

Mergner, H. (1985) Initial recolonization of Funafuti Atoll coral reefs devastated by Hurricane Bebe. *ARB*, **284**, 29 pp.

Mergner, H. and Scheer, G. (1974) The physiographic zonation and the ecological conditions of some South Indian and Ceylon coral reefs. *Proc. 2nd ICRS, Brisbane*, **2**, 3–30.

Mergner, H. and Schuhmacher, H. (1974) Morphologie, Oekologie und Zonierung von Korallenriffen bei Aqaba (Golf von Aqaba, Rotes Meer). *Helgoländer wiss. Meeresuntersch.*, **26**, 238–358.

Michel, A. (1969) Plancton du lagon et des abords extérieurs de l'atoll de Mururoa. *CP*, **13**, 81–132.

Michel, A., Colin, C., Desrosiers, R. and Oudot, C. (1971) Observations sur l'hydrologie et le plancton des abords des passes de l'atoll de Rangiroa. *Cah. ORSTOM, sér. Océanogr.*, **11** (3), 375–402.

Milliman, J. D. (1965) An annotated bibliography of recent papers on corals and coral reefs. *ARB*, **111**, 58 pp.

Milliman, J. D. (1967) The geomorphology and history of Hogsty Reef, a Bahamian atoll. *Bull. Mar. Sci. (Miami),* **17** (3), 519–543.

Milliman, J. D. (1969) Four Southwestern Caribbean atolls: Courtown Cays, Albuquerque Cays, Roncador Bank and Serrana Bank. *ARB*, **129**, 40 pp.

Milliman, J. D. (1973) Caribbean coral reefs. In Jones and Endean editors, *Biology and Geology of coral reefs*, Ac. Press, N. York, vol. 1, 51–92.

Milliman, J. D. and Meade, R. H. (1983) World-wide delivery of river sediment to the Ocean. *J. of Geol.*, **91**, 1–21.

Milliman, J. D. and Supko, P. R. (1968) On the geology of San Andros, Western Caribbean. *Geol. en Mijnb.*, **47** (2), 102–105.

Mitchell, A. H. G. (1968) Raised reef-capped terraces and Plio-Pleistocene sea-level changes, North Malekula, New Hebrides. *J. of Geol.*, **76**, 56–67.

Montaggioni, L. (1974) Coral reefs and quaternary shorelines in the Mascarene Archipelago, Indian Ocean. *Proc. 2nd ICRS, Brisbane*, **2**, 579–593.

Montaggioni, L. (1985) Makatea Island, Tuamotu Archipelago. *5th ICRC, Tahiti*, **1**, 103–158.

Montaggioni, L. and Pirazzoli, P. A. (1984) The significance of exposed coral conglomerates from French Polynesia (Pacific Ocean) as indicators of recent relative sea level changes. *CR*, **3**, 29–42.

Mörner, N. A. (1976) Eustasy and geoid changes. *J. Geol.*, **84**, 123–151.

Mullins, H. T., Keller, G. H., Kofoed, J. W., Lambert, D. N., Stubblefield, W. L. and Warme, J. E. (1982) Geology of the Great Abaco submarine canyon (Blake Plateau): observations from the research submarine 'Alvin'. *Mar. Geol.*, **48**, 239–257.

Mullins, H. T. and Lynts, G. W. (1977) Origin of the Northwestern Bahama Platform: review and reinterpretation. *GSAB*, **88**, 1447–1461.

Munch-Petersen, N. F. (1985) Man and reef in the Maldives. Continuity and change. *5th ICRC, Tahiti*, **2**, 254.

Munk, W. H. and Sargent, M. C. (1948) Adjustment of Bikini Atoll to ocean waves. *US Geol Surv. Prof. Paper*, **260 C**, 275–280.

Munro, J. L. and Williams, D. Mc. B., convenors (1985) Assessment and management of coral reef fisheries: biological, environmental and socio-economic aspects. *Proc. 5th ICRC, Tahiti*, **4**, 543–578 (extensive bibliography).

Muscatine, L., McCloskey, L. R. and Loya, Y. (1985) A comparison of the growth rates of Zooxanthellae and animal tissue in the Red Sea coral *Stylophora pistillata. Proc. 5th ICRC, Tahiti*, **6**, 119–123.

Muzik, K. (1985) Dying coral reefs of the Kyukyu Archipelago, Japan. *Proc. 5th ICRC, Tahiti*, **6**, 483–489.

Nakata, T., Koba, M., Jo, W., Imaizumi, I., Matsumoto, H. and Sugaruma, T. (1979) Holocene marine terraces and seismic crustal movement. *Sci. Rep. Tohoku Univ*, 7th Ser., Geogr., **29**, 195–204.

Nasr, D. H. (1985) Coral reef conservation in Sudan. *Proc. 5th ICRC, Tahiti*, **4**, 243–246.

Nestéroff, W. (1952) Coupes transversales de la Mer Rouge. *C. R. Ac. Sci. Paris*, **235**, 1503–1505.

Nestéroff, W. (1955) Les récifs coralliens du Banc Farsan Nord (Mer Rouge). *Ann. Inst. Océanogr.*, **XXX**, 7–53.

Neumann, A. C. and MacIntyre, I. (1985) Reef response to sea-level rise: keep-up, catch-up, or give-up. *Proc. 5th ICRC, Tahiti*, **3**, 105–110.

Newell, N. D. (1954) Reefs and sedimentary processes of Raroia. *ARB*, **36**, 32 pp.

Newell, N. D. and Bloom, A. L. (1970) The reef flat and the 'Two Meter Estatic Terrace' of some Pacific atolls. *GSAB*, **81**, 1881–1894.

Newell, Imbrie, J., Purdy, E. G. and Thurber, D. L. (1959) Organism communities and bottom facies, Great Bahama Bank. *Bull. Am. Mus. Nat. Hist.*, **117** (4), 181–228.

Newell, N. D. and Rigby, J. K. (1957) Geological studies on the great Bahama Bank. In *Regional Aspects of carbonate deposition. Spec. Publ. Soc. Econ. Paleontol. and Min., Tulsa*, **5**, 15–79.

Nishihita, M. and Yamazato, K. (1974) Human interference with the coral reef community and Acanthaster infestation of Okinawa. *Proc. 2nd ICRS, Brisbane*, **1**, 577–590.

Nougier, J., Vatin-Perignon, N., Cantagrel, J. M. and Cheminée, J. L. (1979) Volcanisme et structure de l'île Mayotte. *C. R. Ac. Sci. Paris*, **D–288**, 211–214.

Nugent, L. E. Jr. (1946) Coral reefs in the Gilbert, Marshall and Caroline Islands. *GSAB*, **57**, 735–779.

Obellianne, J. M. (1958) Contribution à la connaissance géologique de l'archipel des Nouvelles-Hébrides. *Sci. de la Terre* (Nancy), **6**, 139–368.

Oberdorfer, J. A. and Buddemeier, R. W. (1985) Coral reef hydrogeology. *Proc. 5th ICRC, Tahiti*, **3**, 307–312.

Odum, H. T. and Odum, E. P. (1955) Trophic structure and productivity of a windward coral reef community on Eniwetok Atoll. *Ecol. Monogr.*, **25**, 291–320.

Okal, E. A. and Cazenave, A. (1985) A model for the plate tectonic evolution of the East-Central Pacific based on SEASAT investigations. *Earth Planet. Sci. Letters*, **72**, 99–116.

Oliver, J. and McGinnity, P. (1985) Commercial coral collecting on the Great Barrier Reef. *Proc. 5th ICRC, Tahiti*, **5**, 563–568.

Orme, G. R. (1977) The Coral Sea Plateau, a major reef province. In Jones and Endean (eds) *Biology and Geology of Coral Reefs*, Ac. Press, N. York, Vol. IV, Geology 2, 267–306 (with general map).

Orme, G. R. (1978) Reefs of the Great Barrier Reef North of Lizard Island. In

Workshop on the Northern Sector of the Great Barrier Reef, G B R Marine Park Authority, Townsville, 12–43.

Orme, G. R. (1985) The sedimentological importance of *Halimeda* in the development of back reef lithofacies, Northern Great Barrier Reef (Australia). *Proc. 5th ICRC, Tahiti*, **5**, 31–37.

Orme, G. R., Flood, P. G. and Ewart, A. (1974) An investigation of the sediments and physiography of Lady Musgrave Reef. A preliminary account. *Proc. 2nd ICRS, Brisbane*, **2**, 371–386.

Orme, G. R. and Flood, P. G. (1977) The geological history of the Great Barrier Reef: a reappraisal of some aspects in the light of new evidence. *Proc. 3rd ICRS, Miami*, **2**, 37–43.

Orme, G. R., Flood, P. G. and Sargent, G. E. G. (1978a). Sedimentation trends in the lee of outer (ribbon) reefs, Northern Region of the Great Barrier Reef Province. *PTRS*, **A 291**, 85–99.

Orme, G. R., Webb, J. P., Kelland, N. C. and Sargent, G. E. G. (1978b) Aspects of the geological history and structure of the Northern Great Barrier Reef. *PTRS*, **A 291**, 23–35.

Ormond, R. F. G. and Campbell, A. C. (1974) Formation and breakdown of *Acanthaster planci* aggregations in the Red Sea. *Proc. 2nd ICRC, Brisbane*, **1**, 595–619.

Ottmann, F. (1960) Une hypothèse sur l'origine des 'arrecifes' du Nordeste brésilien. *C.R. Somm. Soc. Géol. Fr.*, **7**, 175–176.

Ottmann, F. (1962) 'L'Atol das Rocas' dans l'Atlantique sud tropical. *Rev. Géogr. Phys. Géol. Dynam.*, (2) **V**, 101–107.

Pautot, G. (1983) Les fosses de la Mer Rouge: approche géomorphologique d'un stade initial d'ouverture océanique réalisée à l'aide du Seabeam. *Oceanol. Acta*, **6**, 235–244.

Pautot, G., Guennoc, P., Coutelle, A. and Lyberis, N. (1986) La depression axiale du segment nord de la Mer Rouge (de 25°N à 28°N): nouvelles données géologiques et géophysiques obtenues au cours de la campagne Transmerou 83. *Bull. Soc. Géol. Fr.*, **(8) 11**, 381–399.

Pichon, M. (1972) The coral reefs of Madagascar. In Richard-Vindard and Battistini (eds) *Biogeography and Ecology of Madagascar*, The Hague, Junk, 367–410.

Pichon, M. (1974) Dynamics of benthic communities in the coral reefs of Tulear (Madagascar). *Proc. 2nd ICRS, Brisbane*, **2**, 55–68.

Pichon, M. (1977) Physiography, morphology and ecology of the double barrier reef of North Bohol (Philippines). *Proc. 3rd ICRS, Miami*, **2**, 261–267.

Pirazzoli, P. (1984) Cartographie des hauts fonds par télédétection dans l'archipel des Gambier (Polynésie Française). *L'Espace Géographique*, **3**, 277–284.

Pirazzoli, P. *et al*. (1985) Leeward Islands, French Polynesia. *Proc. 5th ICRC, Tahiti*, **1**, 17–72.

Pirazzoli, P. and Montaggioni, L. F. (1985) Lithospheric deformation in French Polynesia (Pacific Ocean) as deduced from Quaternary shorelines. *Proc. 5th ICRC, Tahiti*, **3**, 195–200.

Pirazzoli, P., Montaggioni, L. F., Delibrias, G., Faure, G. and Salvat, B. (1985) Late Holocene sea-level changes in the Society Islands and in the Northwest Tuamotu Atolls. *Proc. 5th ICRC, Tahiti*, **3**, 131–136.

Pirazzoli, P. and Montaggioni, L. F. (1986) Late Holocene sea-level changes in the Northwest Tuamotu Islands, French Polynesia. *Quat. Res.*, **25**, 350–568.

Popenoe, P., Kohout, F. A. and Manheim, F. T. (1984) Seismic-reflection studies of sinkholes and limestone dissolution features on the Northeastern Florida Shelf. *Proc. 1st Multidisc. Conf. on Sinkholes, Orlando* (Publ. Balkema, Rotterdam and Boston), 43–57.

Purdy, E. G. (1974) Reef configuration: cause and effect. In Laporte (ed) *Reefs in*

Time and Space, Soc. of Econ., Paleont. and Mineral., Tulsa, 9–76 (extensive bibliography).

Purdy, E. G. (1981) Evolution of the Maldive Atolls, Indian Ocean. *Proc. 4th ICRS, Manila*, **1**, 659.

Purser, B. H. and Seibold, E. (1973) The principal environmental factors influencing Holocene sedimentation and diagenesis in the Persian Gulf. In Purser (ed) *The Persian Gulf*, Berlin–Heidelberg–New York, Springer, 1–9 (and separate map showing distribution of coral reefs in the Persian Gulf).

Rabesandratana, H. V. (1985) About some reef utilization in Madagascar. *Proc. 5th ICRC, Tahiti*, **6**, 661–668.

Randall, J. E. (1974) The effect of fishes on coral reefs. *Proc. 2nd ICRS, Brisbane*, **1**, 159–166.

Rathjens, C. and Von Wissmann, H. (1933) Morphologische Probleme im Graben des Roten Meeres. *Peterm. Mitteil.*, **79**, 113–117 and 183–187.

Reed, J. K. (1985) Deepest distribution of Atlantic hermatypic corals discovered in the Bahamas. *Proc. 5th ICRC, Tahiti*, **6**, 249–254.

Roberts, H. H. (1974) Variability of reefs with regard to changes in wave power around an island. *Proc. 2nd ICRS, Brisbane*, **2**, 497–512.

Roberts, H. H. (1983) Shelf margin reef morphology: a clue to major off-shelf sediment transport routes, Grand Cayman Island, West Indies. *ARB*, **263**, 19 pp.

Roberts, H. H. and Murray, S. P. (1983) Controls on reef development and the terrigenous–carbonate interface on a shallow reef, Nicaragua, Central America. *CR*, **2**, 71–80.

Roberts, H. H. and Murray, S. P. (1984) Developing carbonate platforms: Southern Gulf of Suez, Northern Red Sea. *Mar. Geol.*, **59**, 165–185.

Rogers, C. S. (1985) Degradation of Caribbean and Western Atlantic coral reefs and decline of associated fisheries. *Proc. 5th ICRC, Tahiti*, **6**, 491–496.

Rougerie, F. (1979) Caractéristiques générales du milieu liquide lagonaire de l'atoll de Takapoto. *J. Soc. Océanistes*, **62**, 35–45.

Rougerie, F. and Wauthy, B. (1986) Le concept d'endo-upwelling dans le fonctionnement des atolls-oasis. *Oceanol. Acta*, **9**, 133–148.

Rougerie, F., and Wauthy, B. (1988) The atoll functioning by endo-upwelling: implication and generalization of the model to other reef-building formations. *C.R.* (in press).

Russ, G. (1985) Effects of protective management on coral reef fishes in the Central Philippines. *Proc. 5th ICRC, Tahiti*, **4**, 219–224.

Russell, R. J. (1962) Origin of beach rock. *Zeit. Geomorphol.*, **3**, 227–236.

Russo, A. R. (1980) Bioerosion by two rock boring echinoids on Enewetak Atoll, Marshall Islands. *J. Mar. Res.*, **38**, 99–110.

Ruzie, G. and Gachon, A. (1985) Contribution of geophysical techniques in carbonate studies of the atolls. Application to the investigation of the structures of Mururoa Atoll, French Polynesia. *Proc. 5th ICRC, Tahiti*, **6**, 381–388.

Sachet, M. H. (1962) Monographie physique et biologique de l'Ile Clipperton. *Ann. Inst. Océonogr.*, **XL–1**, 1–107.

Sachet, M. H., Stoddart, D. R. and Fosberg, F. R. (eds) (1983) Floristics and ecology of Western Indian Ocean Islands. *ARB*, **273**, 253 pp.

Salomon, J. N. (1980) Les récifs coralliens de Belo-sur-Mer: étude géomorphologique. *Madagascar Rev. de Géogr.*, **37**, 87–109.

Salvat, B. (1979) Recherches sur l'atoll de Takapoto (Tuamotu, Polynésie Française). *J. Soc. Océanistes*, **62**, 5–7.

Salvat, B. (1981) Preservation of coral reefs: scientific studies or economic necessity? Past, present and future. *Proc. 4th ICRS, Manila*, **1**, 225–229.

Salvat, B. and Richard, G. (1985) Takapoto Atoll, Tuamotu Archipelago. *Proc. 5th ICRC, Tahiti*, **1**, 323–378.

Sargent, M. C. and Austin, T. S. (1954) Biologic economy of coral reefs. In *Bikini and nearby atolls. U.S. Geol. Surv. Prof. Paper*, **260 E**, 293–300.

Scatterday, J. W. (1974) Reefs and associated coral assemblages off Bonaire, Netherlands Antilles, and their bearing on Pleistocene and recent reef models. *Proc. 2nd ICRS, Brisbane*, **2**, 85–106.

Scheer, G. (1971) Coral reefs and coral genera in the Red Sea and Indian Ocean. *Symp. Zool. Soc. London*, **28**, 329–367.

Schick, A. P. (1958a) Marine terraces on Tiran Island, Northern Red Sea. *Geogr. Annaler*, **40**, 1, 63–66.

Schick, A. P. (1958b) Tiran: the straits, the island and its terraces. *Israel Explor. J.*, **8** (2), 120–130 and (3), 189–196.

Schlanger, S. O., Garcia, M. O., Keating, B. H., Naughton, J. J., Duncan, R. A., Sager, W. W., Haggerty, J. A. and Philpotts, J. A. (1984) Geology and Geochronology of the Line Islands. *J. Geophys. Res.*, **89**, 11261–11272.

Schmidt, W. (1923) Die Scherms an der Rotmeerkuste von El-Hedschas. *Peterm. Mitteil.*, **69**, 118–121.

Schneider, J. (1976) Biological and inorganic factors in the destruction of limestone coasts. *Contrib. to Sedimentol.*, **6**, 1–112.

Scholten, J. J. (1972) Beach rock: a literature study with special reference to recent literature. *Zentralblatt Geol. Paleontol.*, Teil 1 (9–10), 351–368 and 655–672.

Schroeder, J. H. (1981) Man versus reef in the Sudan: threats, destruction, protection. *Proc. 4th ICRS, Manila*, **1**, 287–291.

Schroeder, J. H. and Nasr, D. H. (1983) The fringing reefs of Port Sudan, Sudan: 1. Morphology, sedimentology, zonation. *Essener Geogr. Arb.*, **6**, 29–44.

Schuhmacher, H. and Mergner, H. (1985) Quantitative Analyse von Korallengemeinschaften des Sanganeb-Atolls, mittleres Rotes Meer. *Helgoländer Meeresuntersch.*, **39**, 375–440.

Schuhmacher, H. and Zibrowius, H. (1985) What is hermatypic? A redefinition of ecological groups in corals and other organisms. *CR*, **4** (1), 1–9.

Scoffin, T. P., Stearn, C. W., Boucher, D., Frydl, P., Hawkins, C. M., Hunter, I. G. and MacGeachy, J. K. (1980) Calcium carbonate budget of a fringing reef on the west coast of Barbados, part 11. *Bull. Mar. Res.*, **30**, 475–508.

Scoffin, T. P., Stoddart, D. R., Tudhope, A. W. and Woodroffe, C. D. (1985) Exposed limestone of Suwarrow Atoll, Northern Cook Islands. *Proc. 5th ICRC, Tahiti*, **3**, 137–140.

Scott, G. A. J. and Rotondo, G. M. (1983a) A model for the development of types of atolls and volcanic islands on the Pacific lithosphere plate. *ARB*, **260**, 33 pp.

Scott, G. A. J. and Rotondo, G. M. (1983b) A model to explain the differences between Pacific Plate Island Atoll types. *CR*, **1**, 139–150.

Scrutton, R. A., Heptonstall, W. B. and Peacock, J. H. (1981) Constraints on the motion of Madagascar with respect to Africa. *Mar. Geol.*, **43**, 1–20.

Sedlak, W. (1977) Some aspects on the stratigraphy and taxonomy of Cambrian fauna found on Lysa Gora (The Swietokrzyskie Mountains, Central Poland). *Mem. BRGM*, **89**, 42–48.

Selected bibliography on the geology of organic reefs (1958). Pacific Sci. Board, Nat. Ac. Sci., Washington DC, mimeogr., 51 pp.

Sestini, J. (1965) Cenozoic stratigraphy and depositional history, Red Sea coast, Sudan. *Am. Ass. Petr. Geol. Bull.*, **49**, 1453–1472.

Shepard, F. P. (1963) *Submarine Geology*, second edition. Harper and Row, New York and London, 555 pp. (first ed. 1948).

Shepard, F. P. (1970) Lagoonal topography of Caroline and Marshall Islands. *GSAB*, **81** 1905–1914 (with map of Truk lagoon).

Slater, R. A. and Phipps, C. V. G. (1977) A preliminary report on the coral reefs of Lord Howe Island and Elizabeth Reef, Australia. *Proc. 3rd ICRS, Miami*, **2**, 313–318.

Smith, S. V. (1978) Coral reef area and contributions of reefs to processes and resources on the World's Oceans. *Nature*, **273**, 5659, 225–226.

Smith, S. V. (1981) The Houtman's Abrolhos Islands: carbon metabolism of coral reefs at high latitude. *Limnol. and Oceanogr.*, **26**, 612–621.

Smith, S. V. and Henderson, R. S. (eds) (1978) Phoenix Islands report. I: An environmental survey of Canton Atoll lagoon. *ARB*, **221**, 183 pp.

Smoot, N. C. (1985) Guyot and seamount morphology and tectonics of the Hawaiian Emperor elbow by multi-beam sonar. *Marine Geol.*, **64**, 203–215.

Soegiarto, A. (1981) The development of a marine park system in Indonesia. *Proc. 4th ICRS, Manila*, **1**, 287–291.

Soegiarto, A. (1985) Coral reef ecosystem in Indonesia: its problems and management. *Proc. 5th ICRC, Tahiti*, **2**, 357.

Steers, J. A. (1929) The Queensland coast and the Great Barrier Reef. *Geogr. J.*, **74**, 232–257, 341–370.

Steers, J. A. (1940) The coral cays of Jamaica. *Geogr. J.*, **95**, 30–42, and **96**, 305–311.

Steers, J. A., Chapman, V. J., Colman, J. and Lofthouse, J. A. (1940) Sand cays and mangroves in Jamaica. *Geogr. J.*, **96**, 312–328.

Steers, J. A. and Stoddart, D. R. (1977) The origin of fringing reefs, barrier reefs, and atolls. In Jones and Endean (eds) *Biology and Geology of Coral Reefs*, Ac. Press, N. York, Vol. 4, Geology 2, 21–57.

Stoddart, D. R. (1962a) Three Caribbean Atolls: Turneffe Islands, Lighthouse Reef, and Glover's Reef, British Honduras. *ARB*, **87**, 151 pp. (and Louisiana State Univ., Baton Rouge, Contrib. 62–3).

Stoddart, D. R. (1962b) Coral Islands by Charles Darwin. *ARB*, **88**, 20 pp.

Stoddart, D. R. (1963) Effects of Hurricane Hattie on the British Honduras reefs and cays, October 30–31, 1961. *ARB*, **95**, 142 pp.

Stoddart, D. R. (1964a) Storm conditions and vegetation in equilibrium of reef islands. *Proc. 9th Conf. Coastal Eng.*, Am. Soc. Civil. Eng., 893–906.

Stoddart, D. R. (1964b) Catastrophic storm effects on the British Honduras reefs and cays. *Nature*, **196**, 4854, 512–515.

Stoddart, D. R. (1965) The shape of atolls. *Mar. Geol.*, **3**, 369–383.

Stoddart, D. R. (1968) The conservation of Aldabra. *Geogr. J.*, **134**, 471–486.

Stoddart, D. R. (1969a) Ecology and morphology of recent coral reefs. *Biological Rev.*, **44**, 433–498.

Stoddart, D. R. (1969b) Reconnaissance geomorphology of Rangiroa Atoll, Tuamotu Archipelago. *ARB*, **125**, 54 pp.

Stoddart, D. R. (1969c) Geomorphology of the Solomon Islands coral reefs. *PTRS*, **B 255**, 355–382.

Stoddart, D. R. (1969d) Geomorphology of the Marovo elevated barrier reef, New Georgia. *PTRS*, **B 255**, 383–402.

Stoddart, D. R. (1969e) Sand cays of Eastern Guadalcanal. *PTRS*, **B 255**, 403–432.

Stoddart, D. R. (1971a) Coral reefs and islands and catastrophic storms. In Steers (ed) *Applied Coastal Geomorphology*, London, Macmillan, 155–197.

Stoddart, D. R. (1971b) Environment and history in Indian Ocean reef morphology. *Symp. Zool. Soc. London*, **28**, 3–38.

Stoddart, D. R. (1972) Catastrophic damage to coral reef communities by earthquake. *Nature*, **239**, 51–52.

Stoddart, D. R. (1973) Coral reefs in the Indian Ocean. In Jones and Endean (eds) *Biology and Geology of Coral Reefs*, Ac. Press, N. York, Vol. 1, 51–92.

Stoddart, D. R. (1975) Sand cays of Tongatapu. *ARB*, **181**, 16 pp.

Stoddart, D. R. and Cann, J. R. (1965) Nature and origin of beach rock. *J. Sed. Petrol.*, **35**, 243–247.

Stoddart, D. R., Coe, M. J. and Fosberg, F. R. (1979) D'Arros and St Joseph, Amirante Islands, Western Indian Ocean. *ARB*, **223**, 48 pp.

Stoddart, D. R., McLean, R. F., Scoffiin, T. P. and Gibbs, P. E. (1978) Forty five years of change on Low Wooded Islands, Great Barrier Reef. *PTRS*, **B 284**, 63–80.
Stoddart, D. R. and Scoffin, T. P. (1979) Micro-atolls: review of form, origin and terminology. *ARB*, **224**, 17 pp.
Stoddart, D. R. and Steers, J. A. (1977) The nature and origin of coral reef islands. In Jones and Endean (eds) *Biology and Geology of Coral Reefs*, Ac. Press, N. York, Vol. 4, Geology 2, 59–105
Stoddart, D. R. and Taylor, J. D., editors (1971) Geography and ecology of Diego Garcia Atoll, Chagos Archipelago. *ARB*, **149**, 237 pp.
Stoddart, D. R., Taylor, J. D., Fosberg, F. R. and Farrow, G. E. (1971) Geomorphology of Aldabra Atoll. *PTRS*, **B 260**, 31–65.
Strasser, A. and Davaud, E. (1985) Recognition of ancient sea levels using sedimentological and diagenetic criteria. *Proc. 5th ICRC, Tahiti*, **3**, 157–162.
Sudara, S. (1981) Conditions of the coral reef in Thailand and the potential usage in the future. *Proc. 4th ICRS, Manila*, **1**, 209–211.
Symonds, P. A. and Davies, P. J. (1985) Origin of the Great Barrier Reef. *Proc. 5th ICRC, Tahiti*, **3**, 153.
Tayama, R. (1952) *Coral Reefs in the South Seas*. Bull. Hydrogr. Office, Maritime Safety Agency, Tokyo, vol. 11, Publ. 941, 3 vols., text 292 pp., two appendices photos and maps.
Taylor, F. W. and Bloom, A. L. (1977) Coral reefs on tectonic blocks, Tonga Island Arc. *Proc. 3rd ICRS, Miami*, **2**, 275–281.
Tectonic Map of the Pacific Segment of the Earth, scale 1 : 10 000 000. 6 sheets, USSR Ac. Sci., Moscow, 1966–1970 (captions in Russian and in English).
Teichert, C. (1947) Contributions to the geology of the Houtman's Abrolhos, Western Australia. *Proc. Linnean Soc. N.S. Wales*, **71** (3–4), 145–196.
Teichert, C. (1958) Cold- and deep-water coral banks. *Am. Ass. Petr. Geol. Bull.*, **42**, 1064–1082.
Teichert, C. and Fairbridge, R. W. (1948) Some coral reefs of the Sahul Shelf. *Geogr. Rev.*, **38**, 222–249.
Ten Brink, U. S. and Watts, A. B. (1985) Seismic stratigraphy of the flexural moat flanking the Hawaiian Islands. *Nature*, **317**, 6040, 421–424.
Thom, B. G. C. (1973) The dilemma of high interstadial sea levels during the last glaciation. *Progr. in Geogr.*, **5**, 170–246.
Thomassin, B. A. and Coudray, J. (1981) Presence of wide hard-ground areas on lagoonal bottoms of the coral reef complex of Noumea (SW New Caledonia). *Proc. 4th ICRS, Manila*, **1**, 511–522.
Tracey, J. I. Jr., Ladd, H. S. and Hoffmeister, J. E. (1948) Reefs of Bikini, Marshall Islands. *GSAB*, **59**, 861–878.
Tracey, J. I. Jr. and Ladd, H. S. (1974) Quaternary history of Eniwetok and Bikini Atolls, Marshall Islands. *Proc. 2nd ICRS, Brisbane*, **2**, 537–550.
Tracey, J. I. Jr., Schlanger, S. O., Stark, J. T., Doan, D. B. and May, H. G. (1964) General geology of Guam. *US Geol. Surv. Prof. Paper*, **403 A**, 404 pp.
Trudgill, S. T. (1976) The marine erosion of limestones on Aldabra Atoll, Indian Ocean. *Zeit Geomorph., N.F., Suppl. Bd.*, **26**, 164–200.
Umbgrove, J. H. F. (1947) Coral reefs of the East Indies. *GSAB*, **58**, 729–777.
Valencia, M. J. (1977) Christmas Island, Pacific Ocean: reconnaissance geologic observations. *ARB*, **197**, 17 pp.
Vallaux, F. (1955) Toponymie et topographie polynésiennes. *Ann. Hydrogr.*, Paris, **1382**, 52 pp.
Van Andel, T. H. and Veevers, J. J. (1967) Morphology and sediments of the Timor Sea. *Bureau of Mines Res., Geol. and Geophys.*, Dept. of Nat. Devel., Commonwealth of Australia, Bull. **63**, 173 pp.

Veeh, H. H. (1966) Th230/U238 and U234/U238 ages of Pleistocene high sea-level stand. *J. Geophys. Res.*, **71**, 3379–3386.

Veron, J. E. N. and Done, T. J. (1979) Corals and coral communities of Lord Howe Island. *Austr. J. Mar. Freshw. Res.*, **30**, 203–236.

Verstappen, H. Th. (1953) *Djakarta Bay, a geomorphological study in shoreline development*. 'S Gravenhage, Trio, 101 pp.

Verstappen, H. Th. (1954) The influence of climatic changes on the formation of coral islands. *Am. J. Sci.*, **252**, 428–435.

Vila, J. M., Butterlin, J., Boisson, D., Feinberg, H., Laclede, J. M. and Pubellier, M. (1986) Evolution sédimentaire et structurale du bassin oligo-miocène de Trois-Rivières (Nord d'Haiti, Grandes Antilles). *R. Géol. Dyn. Géogr. Phys.*, **27**, 183–192.

Vine, P. J., Endean, R., Stablum, W., Devaney, D. M., Randall, J. E., Marsh, J. A. Jr. and Tsuda, R. T. (1973) Crown of Thorns (*Acanthaster planci*) plagues. *ARB*, **166–170**, five papers in same volume, 178 pp.

Von Arx, W. S. (1954) Circulation systems of Bikini and Rongelap lagoons. *US Geol. Surv. Prof. Paper*, **260 B**, 267–273.

Von Wissmann, H. (1954) Der Karst der humiden Heissen und Sommerheissen Gebiete Ostasiens. *Erdkunde*, **8**, 122–130.

Weber, J. N. (1974) 13/C–12/C ratios as natural isotopic tracers elucidating calcification processes in reef-building and non-reef building corals. *Proc. 2nd ICRS, Brisbane*, **2**, 289–298.

Wells, S. M. (1981) International trade of ornamental corals and shells. *Proc. 4th ICRS, Manila*, **1**, 323–330.

Weydert, P. (1968) Zonation des ensembles récifaux internes de la baie de Tuléar (Sud-Ouest de Madagascar). *C. R. Ac. Sci. Paris*, **266**, 1367–1369.

Weydert, P. (1973) *Morphologie et sédimentologie des formations récifales de la région de Tuléar (Sud-Ouest de Madagascar)*. Thesis Aix-Marseilles, 646 pp.

Weydert, P. (1974) Morphologie et sédimentologie de la pente externe de la partie nord du Grand Récif de Tuléar (Sud-Ouest de Madagascar). *Mar. Geol.*, **17**, 299–337.

Weydert, P. (1979) Direction de croissance des éperons sur le front d'un récif-barrière: l'exemple du Grand Récif de Tuléar (Madagascar). *Mar. Geol.*, **30**, M9–M19.

Wiens, H. J. (1962) *Atoll Environment and Ecology*. Yale Univ. Press, New Haven and London, 532 pp.

Wood-Jones, F. (1912) *Coral and atolls: a history and description of the Keeling-Cocos Islands*. London, L. Reeve & Co., 392 pp.

Woodley, S. J. (1985) The Great Barrier Reef Marine Park: the management challenge. *Proc. 5th ICRC, Tahiti*, **4**, 259–264.

Wu, B. L. (1985) Coral reefs of Hainan Island in the South China Sea. *Proc. 5th ICRC, Tahiti*, **2**, 412.

Yabe, H. (1942) Problems of the coral reefs. *Tohoku Imp. Univ., Geol. and Paleo. Inst., Rept.*, **39**, 1–6 (in Japanese; see Burke 1951 for English summary).

Yonekura, N. (1975) Quaternary tectonic movements in the outer arc of Southwest Japan with special reference to seismic crustal deformation. *Bull. Dep. Geogr. Univ. Tokyo*, **7**, 19–71.

Yonge, C. M. (1930–1931) Studies on the physiology of corals. *Sci. Rep. Great Barrier Reef Exped.*, **1**, 13–92.

Yonge, C. M. (1972) Aspects of productivity in coral reefs. *Proc. 1st ICRS, Mandapam*, 1–12.

Zamarreno, I. and Debrenne, F. (1977) Sédimentologie et biologie des constructions organogènes du Cambrien Inférieur du Sud de l'Espagne. *Mem. BRGM*, **89**, 49–61.

Zucker, W. H. (1985) Reef management in the Pacific Region. *Proc. 5th ICRC*, Tahiti, **2**, 424.

Location Index